The Boy from Bear Swamp

Library of Congress Card Number: 00-109027
ISBN: 0-9675233-2-X

First Printing, December 2000
Second Printing, March 2001

Cover design by Steve McAfee
Book design by Nancy Hutchison Prentice

Printed in the United States of America

04 03 02 01 10 9 8 7 6 5 4 3 2

Bocage Books
4210 Abington Walk, Kennesaw, Georgia 30144
bocagebooks@mindspring.com

The Boy from Bear Swamp

William Clyde Stegall

...From the protracted revival meetings near Bear Swamp, when I was a child, to the present day, the spiritual journey has been one of learning, of faith, and of assurance that, no matter how rugged the course, we are never alone....

Stegall

(1921-2000)

The Boy from Bear Swamp

Portrait of William Clyde Stegall

Chapter 1

The Beginning

I was born on Sunday morning, December 4, 1921, in a little log cabin along Coneross Creek three or four miles upstream from Walhalla, South Carolina, with Dr. Thode in attendance.

My name, William Clyde Stegall, was duly recorded in the family Bible, as the son of Berry C. and Weedie White Stegall.

Years later, when I applied for my first passport, I discovered that my original birth certificate showed that a Stegall child had been born on December 5, but that child had been registered as a female. So to obtain an affidavit, stating that this six foot, six inch World War II veteran was a male, my mother got out the old family Bible and, together, we went to the county court house in Walhalla to straighten out the mistake. I then received my passport.

A lot of memories have paved the road from that isolated log cabin where I was born, to the present day. Like so many who grew up in the rural South in the Great Depression, I have known poverty, disappointment, and sorrow. But I have also experienced the richness of faith, family, and friends. And it is because of the richness of the journey that has taken me to places I never dreamed of, and the friends and family who have made my life meaningful, that I wish to leave behind this last remembrance.

I was an only child, and because we never lived near any children when I was growing up, I was often lonely. I could hardly wait for school to start each year so that I would have someone to play with. Years later, when I asked my mother why they didn't have more children, she said, "We could not afford it."

The Stegall family had not always been poor. In earlier days, many had been landowners and prosperous people. But the Civil War changed the lives of many Southern families, including our own. The personal stories from that era became our own generation's family history, told and retold, so that we would remember not only who we were, but also the enormous cost to later generations when a nation is divided.

The Stegall story begins, not with my parents and me, but with my Civil War grandfather, William "Pap" Stegall, and his father, before him.

Chapter 2

William "Pap" Stegall

I was only three years old when my grandfather, William "Pap" Stegall, died at the age of ninety. The only thing about him that I can clearly remember was Pap's long white beard, as he sat in a chair on the back porch at Uncle Doc's house in the mill village in Walhalla.

He was a big man, six feet, four inches tall. But by that time, in late December of 1924, he was so frail that someone had to help him from the bed to the porch. My grandfather died a few hours after the new year began on January 1, 1925.

Pap had been a Civil War soldier, serving in the Sixth South Carolina Cavalry of the Confederate Army. On May 16, 1861, he had married a young widow, Mary Ann Cargill Montgomery, whose late husband, Homer

Montgomery, had left her with two young children, John and Mary Elizabeth, aged two and four. The Montgomery-Stegall marriage notice, printed in the *Keowee Courier*, stated that the two, both from Pickens, South Carolina, were married on Thursday, May 16, 1861, by Rev. C. McKendree Smith.

Their first child, Thomas Sullivan Stegall, was born a year later, on June 9, 1862, and named for one of Pap's good friends, Mims Sullivan, who was his banker in Greenville, S.C. When the baby was one month old, Pap entered the Confederate Army. The family was never sure whether he had been conscripted, or drafted, or whether he volunteered. I suspect it was the latter, and it was his good friend, Sullivan, who more than likely played a major role.

Capt. Mims Sullivan, who had some university, as well as military, training, had probably been entrusted to create a company of cavalry to report to Camp Preston in Columbia. Naturally, he would invite his friends and customers to become part of that company.

So Pap, who was a rather successful farmer and landowner, rode with Sullivan and the newly formed Company A, toward Columbia and Fort Preston. The first muster-in roll available is dated July 10, 1862. He shows up as being a part of more than one military unit, although it is obvious that they were the same unit with different names.

A cavalry company (or troop) consisted of one hundred men, and a regiment was made up of ten companies, to form a cavalry of a thousand. But oftentimes, the units had less than the optimal number. Also, unlike the Union soldiers, who were issued horses and saddles by the government, each Confederate trooper furnished his own personal horse and saddle.

Officers under whom William "Pap" Stegall served, according to family tradition and affirmed by *Butler and his Cavalry, 1861-1865*, by U.R. Brooks, were General Wade Hampton, General M.C. Butler, Col. Hugh Aiken, and Lt. Col. L.P. Miller. The officers of Company A were Capt. M.A. Sullivan, 1st Lt. Harvey Woods, 2nd Lt. J. Mims Sullivan, and 3rd Lt. John M. Sullivan.

Mims was a familiar name, as well as *Sullivan*, in Company A. The namesakes were possibly fathers and sons, brothers, uncles, and cousins.

Before the regiment was fully organized, the men were ordered around the middle of July 1862 to the coast of South Carolina, to guard the Savannah-Charleston Railroad. Adams Run, across the sound from Hilton Head Island, became their headquarters. For the next twenty-one months, until April 1864, the Sixth South Carolina Cavalry was engaged in many small battles and skirmishes up and down the coast.

There is a book in my Civil War collection that gives an excellent account of the part that Hilton Head Island played in the war. The book, *Hilton Head Island in the Civil War*, by Robert Carse, was given to my wife Ann and me by Marge and Stan Smith of tennis fame, when we stayed in their condo on our visit in 1979.

Antebellum Hilton Head Island was owned by Southerners, who raised long staple cotton and indigo on the plantations. Since the island was isolated and could be reached only by boat, most of the landowners lived on the mainland, with the plantations run by overseers and blacks. The few landowners who lived there year round fled during the Yankee invasion.

The Union Army had up to 65,000 men stationed on Hilton Head Island. It was supposed to have been a staging area, but from what I read about the times, it seems that the island became more of a vacation spot for the soldiers. They had all the comforts of home there—barracks, stores, barbershops, and two newspapers.

The Confederates never had more than 4,000 men guarding the Savannah-Charleston Railroad. The Union forces made several feeble attempts to cut this line, but they were always driven back by the Confederates. It is my opinion that the Yankees found living quite pleasant on Hilton Head Island and were in no hurry to give up the sunshine and seashore. They were of sufficient strength to march into Atlanta during the early part of the war, which, perhaps, would have ended the conflict in that first year.

Hilton Head Island was also used to recruit and train black troops, even before the U.S. War Department gave its official stamp of approval.

One of the stories passed down by Pap Stegall was the account of a fierce battle near Charleston where black troops were used by the Union to storm the fortifications that the Confederates were using. Vastly outnumbered, Pap and the other Confederate soldiers ran out of ammunition as wave upon wave attacked. When the troops came over the walls of the fortification, the Confederates then resorted to using their rifle barrels as weapons, winning the battle in hand-to-hand combat. I have not been able to document this, but the battle, as described by Pap, could easily have been the Battle of Fort Wagner, taking place from July 18-September 7, 1863. Black troops were used there by the Union forces for only the fourth time in the war. They suffered enormous casualties, and the story of their unit, the 54th Massachusetts Infantry, became the subject of the 1989 film, *Glory.*

By the 1930s, during which our Future Farmers of America Chapter in Walhalla camped several times at Bluffton, S.C., we could look across the bay and see Hilton Head Island. Then owned by rich Northerners and used as a hunting preserve, the island was mostly forest, with some cultivation of long staple, or Sea Island, cotton. At that time, the only way to the island was still by boat.

Today, both Bluffton, which in the 1930s was unincorporated and had no electricity, and Hilton Head Island are the playgrounds of the rich and retired, with access from one to the other by an impressive causeway, as well as the Intracoastal Waterway. Yet, there is little evidence of the Civil War historical sites, except on the north end of the island where a few concrete foundations are the only remains of the heavy gun emplacements.

At Adams Run, on January 6, 1864, Pap requested

a leave of absence from his company for fifteen days to bring a new horse to camp. The signed permission contained a detailed description of Stegall, with the notation that if he were not back by a certain date, January 21, he would be considered a deserter. This indicates that Sullivan thought Stegall might overstay his leave, since he had a wife, three children, and a farm that was probably deteriorating.

In those days, it would have taken a man a week to travel from Adams Run near Charleston to the Greenville upcountry area and, of course, a week to return. That would leave only one day at home, under the best of circumstances. The records state that Stegall was returned to camp, under guard, on February 5. Knowing the friendly relationship existing between Sullivan and Stegall, I assume that Sullivan was looking after Stegall's best interest by sending someone after him. If the guard had left on January 22, the day after Pap had been due to report back to headquarters, he would barely have time to travel to Greenville, find Pap, and return by February 5. There is no record of disciplinary action taken against Pap. Sending for him before he got into serious trouble indicates that Sullivan was aware that Pap would more than likely overstay his leave.

In April 1864, Pap and the 6th Regiment of the South Carolina Cavalry left the South Carolina coast and, along with the 4th and 5th regiments, became part of Butler's Brigade (organized on May 18, 1864) in northern Virginia, where some of the fiercest fighting of the war occurred. It doesn't seem possible that so many could have survived the battles. Yet, there were few casualties in Pap's company.

The book, *Butler's Cavalry, 1861-1865*, states that the Sixth South Carolina Cavalry was a wild bunch when serving on the S.C. coast. But when Lt. Col. L.P. Miller took charge of the regiment, the unit became one of the most disciplined regiments in the army. That is not surprising, since Lt. Col. Miller had the reputation of being one of the finest disciplinarians in the entire Confederate Army.

In Virginia, they were armed with swords, pistols, and Enfield rifles. The Enfield was breech-loaded instead of muzzle-loaded, thus making the firing much faster. Because of the terrain, the men did much of their fighting while dismounted, although they could move swiftly when the situation warranted it.

One of the bloodiest cavalry battles of the entire

war was the Battle of Trevilian Station on June 11-12, 1864. General Sheridan had driven a wedge between Gen. Wade Hampton's cavalry division and that of Fitzhugh Lee. But once again, the cavalry dismounted and fought on foot. The Confederate victory kept General Sheridan from reaching Charlottesville and linking up with the Union army in the valley.

Then, there was the famous Cattle Raid, when many heads of cattle were captured from the Yankees and driven back through the Virginia swamps into Confederate territory. Although I am not absolutely sure, I believe that Pap took part in that assignment, too.

In January 1865, Butler's Brigade was ordered back to the Carolinas, to get fresh horses and to help in heading off Sherman's march through the South. From the middle of February until the close of the war in late April, this regiment of cavalry was in daily and nightly contact with General Sherman's forces, in many hit and run battles between Columbia, S.C. and Greensboro, N.C.

After General Lee's surrender at Appomattox on April 9, 1865, this regiment, assigned to the Army of Tennessee under Gen. Joe Johnson, continued fighting against Sherman until the end of April. On April 23,

General Joe Johnson and Sherman met at the Bennett house near Durham, North Carolina, and for several days negotiated the surrender. The final surrender came on April 26, and the men were separated to go home on May 1, 1865.

Actually, the Sixth Cavalry never surrendered. On the eve of the surrender of Joe Johnson's army, the regiment marched all night and, on the next day, they disbanded at the North Carolina-South Carolina state line, returning to their homes with all of their arms and equipment. At that time, the regiment consisted of only 150 mounted men.

Another story handed down through the family questions whether Pap disbanded with the others of his company at the state line. Pap had contracted pneumonia, or a similar respiratory illness, and he was given leave to go home and recuperate. One day, he was lying down, soaking up the sun on the backside of the barn, with his Enfield rifle beside him.

Either an enemy patrol, or a band of carpet-baggers, arrived at the house with the intent of taking away anything of value and burning down the house. Pap was prepared to defend his family and home. But the group, after looking the situation over, took no action. Perhaps sympathetic toward this supposed widow, her children, and her unattended farm, they left peaceably.

After the war, Pap Stegall and his friend, Mims Sullivan, who were disenfranchised as all former Confederate soldiers were, returned to family life. But the Reconstruction period was a difficult one, and Sullivan's troops had one more battle to fight. Sullivan, who had gone back to work at the Greenville bank, was accused of killing a carpetbagger named Gilrease. Sullivan was placed in jail and, with the carpetbaggers and scalawags in control of the government, his future looked dim. Many of Mims Sullivan's former troopers, including my grandfather, got together and rode their cavalry mounts into Greenville and forced the release of Sullivan. After that, nothing more was done toward prosecuting Sullivan.

Pap and his wife, Mary Ann, had five more children, who lived into adulthood—Walter, James, Mrs. J.R. Hamilton, Mrs. T.C. Couch, and Mrs. Sug Ellison. Three died in infancy.

Pap worked hard to support his family and he survived Reconstruction without being completely destroyed. He had a big farming operation near the Anderson-

Greenville county lines. And he owned and operated a cotton gin, which was powered by water from a millpond on a nearby creek.

James Couch, the husband of Ruby Couch, showed me the remains of an old canal believed to have brought water from a creek to the cotton gin, which was located somewhere between Mt. Pisgah Baptist Church and one of the old Stegall home places.

Since Pap was a hands-on type of person, he actually ran the ginning operation. But he hired a man, a relative, as bookkeeper to keep an accurate account of all the records and money.

After several years of a successful operation, the office building and the gin mysteriously burned to the ground. All of the records and all of the money were either destroyed or they vanished. The bookkeeper vanished also.

Years later, the former bookkeeper returned to this South Carolina community after having made a fortune in Texas. He became a respected citizen in the eyes of many people. At his gravesite is a life-sized statue, a tribute to his worldly success.

Sometime before the cotton gin episode and the vanishing of the money and records, Pap Stegall became ill with what was diagnosed as consumption. Since he had done much of the ginning, and had come in contact with the lint floating through the air, that was thought to have been the cause of the disease. In those days, consumption was a death sentence, with the patient given only a few months to live.

When this happened, Pap was ill-advised to transfer all his assets and property to his wife's name. But his wife, Mary Ann, died instead. According to South Carolina law, Pap received only a child's portion, or one-seventh of the estate. I am not sure how this figure was arrived at, since it does not coincide with the number of children. Mary Ann and the two children by her first marriage had inherited much land and other assets from the estate of Homer Montgomery. Dividing his own assets with all the children was undoubtedly a severe financial blow for Pap Stegall.

Pap was the oldest son of Spencer Stegall, who had moved from the area to Oconee County.

In traveling from his home in the Easley-Williamston community to visit his father, Pap had to cross the Seneca (Keowee) River. Since there was no bridge at the time, the only way to get across was by a horse-drawn ferry. Cherry's Crossing was just below where the Seneca bridge near Clemson is today. The ferry was operated by a man named Williams, and his helper was his sister, Sue Williams, whose mother was a Crenshaw.

As Pap used this ferry a number of times, a friendship with Sue developed. The two were married in 1891, when Sue was thirty-six years old. Pap and Sue had four children—Ada, 1892, William Spencer (Bub), 1894, Berry Clyde (my father), 1897, and Edward Clay (Doc), 1899. Shortly before Grandmother Sue passed away in 1920, at the age of sixty-five, my mother and father were married.

Pap lived for five more years. He died at 3 A.M. on January 1, 1925. His funeral notice states that he was ninety-one, but according to family records, he was ninety.

His obituary appeared in the *Keowee Courier* on January 7, 1925, under the headlines:

William Stegall, Confederate Veteran, Passes On

We always hope and expect to have joy on the New Year's morn, but this New Year brought sadness to the home of E.C. Stegall, of Walhalla, for it was at 3 o'clock in the morning of the first day of the New Year that his father, William Stegall, passed on to the great beyond.

William Stegall was born ninety-one years ago in Oconee County. He was a son of the late Spencer Stegall. He was twice married. His first wife was a Mrs. Mary Montgomery, who preceded him to the grave thirty-six years ago. To this union, six children were born—Thomas Stegall of Anderson County, Walter Stegall of Easley, James Stegall of North Carolina, Mr. (Mrs.?) J.R. Hamilton of Easley, Mrs. T.C. Couch of Liberty, and Mrs. Sug Ellison of Piedmont. Mr. Stegall is also survived by two brothers and two sisters—James and Pad Stegall of Greenville County, and Mrs. Ann Jameson of Pelzer, and Mrs. Melinda Day of Anderson.

Mr. Stegall's last wife was a Miss Sue Williams, who passed away about five years ago. To this union, four children were born—E.C., W.S., B.C. Stegall, and Mrs. Charles Smith, the latter preceding her father to the grave about one year ago. His grand and great-grandchildren will number about 100 or more.

Mr. Stegall was a Confederate soldier. He served throughout the war, and was a brave and gallant soldier, always performing his duties to the letter.

The deceased was a farmer by occupation, and followed the tilling of the soil all his life. He at one time owned one of the finest farms in Greenville County, but after losing many fortunes, he moved back to Oconee and remained in this county until his death.

Mr. Stegall was a kindhearted and righteous man, always helping the poor and needy whenever he could. He always served God the best he knew how and only a while before his passing away he said he was going to heaven.

For the last five years he had made his home with his children. He was at the time of his death living with his youngest son, E.C. Stegall, and for the last two years had been almost totally helpless. He was very ill for three weeks before his death came to relieve his sufferings. His remains were laid to rest in Rock Springs cemetery. He was a member of Rock Springs church, which church he had helped to build when he was quite a young man. Rev. Hyatt preached the funeral, and it was he who baptized the deceased when he was but a mere child.

All who knew Mr. Stegall can truly say that a good man has gone to his reward, which was justly earned, and the following lines are peculiarly appropriate in regard to the bearing his death had on his home—

"A chair is vacant in our home,
A voice we loved is stilled—
A place is vacant in our home
Which never can be filled."

That is the life, as I know it, of my grandfather, William "Pap" Stegall.

(Obituary reprinted by permission of Keowee Courier)

William "Pap" Stegall with family
Oconee County, circa 1905.

Chapter 3

Spencer Stegall

My dad, Berry, said that my great-grandfather, Spencer Scorcher Stegall, and all eight of his sons, including Pap, served in the Civil War.

A muster roll, dated April 11, 1863, shows a Spencer Stegall as a member of Company Number 4, First Battalion, Fifth Regiment, South Carolina Militia. He was listed as fifty-two years old. There is a discrepancy in the age given on his cemetery marker at Rock Springs Baptist Church near Easley, S.C., which lists his birth date as 1818. That would have made him only about sixteen or seventeen when Pap was born.

I am inclined to believe that he was born in 1815, which would agree with the 1870 census and make him nineteen when his first son, Pap, was born. This would also be closer to the age listed on the muster roll of 1863.

Still, there are a number of inconsistencies in the war records. When Pap applied for a pension from the state of South Carolina, there were quite a few errors in the application, including the date when he began service in the Confederate Army and the date of his discharge.. But by that time, Pap was old, and the people helping him to qualify did not have all the facts at hand. They were simply trying to help an old war veteran.

So, the facts surrounding Great-Grandpa Spencer probably contain a few mistakes, as well.

One of his sons, a brother to Pap, was George Washington Stegall, who was captured at Stoney Creek, Virginia on December 1, 1864. He was released on June 19, 1865, after the war. Spencer also had a younger brother, William D. Stegall, who was evidently wounded in battle and received treatment at Jackson Hospital in Richmond, Virginia. His outfit was listed as Company D, Sixth Regiment, South Carolina Cavalry.

In searching through the archives in Columbia, I found a number of Stegalls serving from several states, not only from the South, but also from the North.

But it is what happened to Great-Grandpa Spencer after the war that provided the family story of his grown sons taking him behind the barn and giving him a good thrashing.

At some point later in his life, as he was growing old and he could barely see, Spencer needed extra care. His wife had probably died by then, so he hired a housekeeper, a young white woman.

Shortly thereafter, the young woman became pregnant with his child. This upset Spencer's sons in the worst possible way. I am not sure whether Pap, his oldest, was involved with the others. But the rest of his grown sons took their father behind the barn and gave him a good whipping for his conduct.

This was such an embarrassment to the whole community, and his sons gave him such a hard time, that he moved to Oconee County. The young woman apparently continued to live with the old gentleman for a time, and a son was born. When she left, it was reported that she had lined her suitcases with all the cash that the old man had.

Spencer Stegall, considered to be the patriarch of the Stegall clan, owned much property and had been highly thought of as one of the most outstanding citizens of the Greenville/Anderson community, with a number of outstanding children. His miscue in later life was a shock for everyone concerned.

As a child, I recall travelling along the road from Bear Swamp to the Richland community. I would pass by the old home place, where apparently that son lived most of his adult life. He, too, had a son, who became one of my father's good friends and neighbors.

"Pap's" second wife, Sue, Berry's mother

Chapter 4

WEEDIE AND BERRY

My mother, Weedie White, was the oldest child of Wesley and Tina White. Her mother, Tina, was either one-fourth or one-half Cherokee. Born on March 27, 1900, Weedie helped raise her six younger brothers and sisters, since her mother worked alongside the men in the fields.

On October 19, 1919, when she was nineteen years old, Weedie, and my father, Berry, were married. Living at first with his parents, Pap and Sue, they moved in 1921 to the log cabin where I was born. This was also the same log house where my grandfather Wesley White had been born in 1857. The cabin was later moved to a farm owned by W.O. and Aunt Sally White.

From Uncle Will's old place, to Grandfather White's homestead, to the Gambrell place in Bear Swamp, to Earnest Smith's place, and back to the Gambrell place, the family moved every year until I started to school.

I can still remember the wagon ride when we first moved in 1925 to the Gambrell place. That was the same year that Pap died. But the year I remembered as the worst was 1926. I was sickly that year. And my first dog, Pood, and my dad's dog, Cliff, had to be shot because both had been bitten by another dog that had rabies.

But then, 1927 came. We moved back to the Gambrell place, which was located close to both the Gambrells and the DuBoses, and I started to Blue Ridge School. Up until then, as an only child, my only playmates had been my cousins, whom I saw every few weeks.

Even though I was lonely as a small child, I was secure in my parents' love. My mother always said that the happiest day of their lives was when I was born. And she showed that love in hugs and kisses. As I grew older, though, I rebelled against this outward display of affection.

My father was different. He never once actually told me that he loved me. But he showed it in a thousand ways. Except when I was at school, I spent virtually the entire day with my father, helping him with whatever chore he was doing. Consequently, I learned to do most of the things he did.

In my opinion, my dad could do anything. And I was not the only one who thought so. He was heavily in demand by all his neighbors to help them with their crops. He had the reputation of laying off the straightest planting rows of anyone in the community. He could do all sorts of field work—plowing, breaking in horses, cutting timber, sharpening saws, axes, plows, and other farm implements, milking cows, resoling shoes, using his own last, cutting crossties, re-rimming wagon wheels, making new seats for chairs out of white oak splints, and doing all kinds of blacksmith work, such as shoeing mules.

From the time I could walk, my father took me with him wherever he went. I remember especially, that when we took our cotton to the gin, we always stopped and got a hot dog at Zeigler's Café.

After I started to school, my dad and I would set out rabbit traps or boxes in the winter. The only way to get back and forth to school was to walk. So on the way home each day, I would go by and collect the rabbits.

Occasionally, we would sell a cleaned rabbit to our

friends in Walhalla for ten or fifteen cents, since rabbit meat was good eating. It tasted very much like chicken.

One afternoon, I found a huge and very beautiful skunk in one of the boxes. Needless to say, I did not try to remove the skunk. When our neighbor, Cantey DuBose, found out about it, he sent for his brother, an amateur taxidermist in Seneca. He put the skunk to sleep with ether, and then took it back to Seneca and mounted it. It was beautiful. DuBose gave me fifty cents for the skunk, which was almost a day's work in those Depression days.

When school was out, baseball was our favorite sport in the summer. One Sunday afternoon, we had a baseball game between two pickup teams in the community. On this occasion, my dad was the catcher. He had a mitt, but not a breast protector nor facemask. A hard ball from the pitcher was fouled by the batter and it hit my dad right below the eye, which broke his cheekbone. From that time on, he had a sunken place under his left eye.

He was six feet tall, and back in those days, he was considered a tall person. He wore a size seven shoe and weighed about one hundred and sixty to one hundred and eighty pounds. When he was fifteen, he had an accident, almost chopping off a foot with an ax when he was cutting wood. Because of this, he did not serve in World War I.

According to my mother, Berry was not a healthy child. There were many days when he was not able to work in the fields with the rest of his family. Born March 7, 1897, at Richland, in Oconee County, he had asthma, which continued to plague him all his life. Although I was not aware of it, my mother and father both prayed that he would live long enough to see me finish high school in 1938. This he did, as well as being present at my graduation from Clemson College.

I can recall, when I was about fourteen, and we were living at the old Addis place, I was sure that my father was dying. He was running a high fever, coughing, and having a terrible time breathing. I got on our old horse, Maude, and rode bareback into West Union to get Dr. Wycliffe to come. He was not there, so I left word that my father was in bad shape and really needed his services. It was not long before Dr. Wycliffe showed up in his old A-Model Ford. He gave my dad the right kind of medicine, apparently, because he began to improve immediately.

The only thing that we had to pay Dr. Wycliffe with was a smoked ham. We also had a smoked shoulder, which was not considered as nice. Instead of the finer smoked ham, Dr. Wycliffe chose the smoked shoulder for payment. Of course, he never made a great deal of money, but I'm sure that he received a greater reward because he spent his entire life taking care of people like my dad.

My dad loved people and he had many friends who loved him. Our house, although the most modest one in the community, was always the gathering place for the men to talk, play cards, and rehash the world's ills.

Dad was also very intelligent and a good manager. Some of the time, as a tenant farmer, he worked on halves and at other times, on thirds. The last several years that he farmed, he rented the property for a set annual fee. But when he had a plowing accident that broke his leg in 1939, that put an end to his farming. Years earlier, he had been the community barber. So once he got well, he rented an existing barbershop in West Union from Miss Murphree, the wife of Sam Albertson. He opened the shop every Friday and Saturday from 1940 until his death, making more money in those eight years than he had ever made in farming. And mostly to help out his friend, he also occasionally worked at Albertson's service station and store.

In 1945, he bought a house in the city of Walhalla, but only lived there a short time. He then bought about seven acres on the north side of Walhalla on the road leading to Highlands, N.C.

My dad liked to hunt and was an excellent shot with shotgun and rifle. He could train hound dogs to hunt rabbits, possums, coons, and birds. I learned that he and the Baptist preacher at Rocky Knoll Baptist Church became good friends because they both enjoyed hunting.

The last time I saw my dad was on February 1, 1948, when my wife Ann and I moved north. The two of us had just spent the day in New York City, our first visit. When we returned to our little one-room apartment in New Jersey, on June 19, 1948, we found a telegram from a cousin, waiting for us. "Your dad is dead." This was the greatest loss I have ever experienced, for he was the best dad a son ever had.

He set an excellent example for me. He was always truthful, dependable, and he never gave up until the job was done. I never remember his using a curse word. But he smoked, and I believe that this, combined with his asthma, contributed to his short life.

His biggest fault was that he would get very upset if someone questioned his word. He had many contracts with landowners and merchants, but I never remember his having to sign a contract, or requiring someone else to sign a contract.

He was buried at Coneross Baptist Church cemetery. Although there is no record that he ever actually joined a local church, my dad was a fine example of a Christian. He often took the family to services at both Coneross and Poplar Springs Baptist churches. And I like to think that during those times of hunting together, my dad and the Baptist preacher discussed the Gospel.

For fifty-one years, he always thanked the Lord for letting him live that long in his weakened condition. And each day, he gave thanks for his beloved Weedie.

My mother lived thirty-five years longer than my father. During much of this time, she was a school lunchroom worker, beginning in 1936 at Blue Ridge School, and then later in Walhalla. In 1932, she had gone to Highlands, N.C. in an attempt to make some extra cash. But things evidently did not work out, because she stayed only a month.

Prior to 1936, Blue Ridge had no school lunch program. Since Granny White lived across the field from the school, I would walk there at "Big Recess" and Granny would prepare lunch for me. When the school lunch program was begun, after I had graduated from grammar school, my mother was hired at eighty cents a day. One mother came each day as a volunteer to help her. Each child brought from home a bowl, a spoon, and a piece of bread. Then Blue Ridge School became the first school in the county to begin baking bread. The government supplied foodstuffs, including canned and cooked roast beef, Washington State apples, creamery butter, and cheese. When Blue Ridge was closed, she started working in the Walhalla school system, and continued until she was seventy-five. She was greatly loved by everyone and, even to this day, I see people in Walhalla who tell me what a fine person my mother was. She was a great lady in many respects.

Her biggest fault, perhaps, was in making up her mind on a matter and not changing it, even though all logic pointed in the opposite direction.

In all those years that she was a widow, I never knew of my mother's having but one special male friend, a well-respected old gentleman who lived on the north side of Walhalla, not far from the housing project in which she lived. They were very congenial and were perhaps engaged at one time. But it seemed there was a problem with the old gentleman's children, who might not have wanted to share any inheritance. This did not seem to concern my mother unduly. They still remained friends. And she had many friends from the old days, who lived in the same housing project. It was an unusually good community, so different from the typical housing project in a big city.

My mother never weighed more than one hundred and five pounds her entire life. And as she grew older, she probably weighed less. I finally convinced her to move to Birmingham to be closer to her family, that included four grandchildren and dozens of great-grandchildren

The last four or five years of her life were spent at St. Martin in the Pines Retirement Center, sponsored by the Episcopal Church. There, she died of a brain tumor on July 9, 1986.

My good friend, Lloyd Morgan, handled the funeral arrangements. Weedie White Stegall was buried beside her loved ones, including her husband Berry, in the cemetery at Coneross Baptist Church, between Walhalla and Westminster, South Carolina.

Berry Stegall, 1918

Weedie Stegall, 1948

Alonzo, Buddy, Weedie, with Thode and Inez White
June 15, 1929

William Clyde Stegall with mother Weedie, Stumphouse Mountain Road, 1929

Berry Stegal, 1918

Chapter 5

THE GREAT DEPRESSION

We did not know we were ever in a depression until we read about it after World War II. For us, there was no let up from 1926, the first year that I remember, to 1935, the year we moved to the old Addis place.

During the ten years from 1925 to 1935, my dad was a sharecropper. He furnished the labor for farming and paid for one-half of the fertilizer and seed. In return, he received one-half of the crops. The money crop was cotton, but we also grew lots of corn. The landowner, of course, provided the land, furnished the horses or mules, provided all the tools and equipment, and paid for the other half of the fertilizer and seed.

In looking back, I really do not see how my father managed to provide for his family. Each spring he would go to the bank or to some individual, often the person who owned the supply house, and he would borrow anywhere from one hundred to three hundred dollars on which to "make a crop." He used the money to buy his share of fertilizer and seed, and often had a few dollars left over to buy staple groceries that we could not raise on the farm.

At harvest time, when Dad sold the cotton, we would first pay off the debt that he had borrowed back in the spring. This debt had to be paid before we could even think about any money to buy clothes or shoes or staple groceries for the winter.

During this time, cotton sold for ten cents a pound, but I can recall one year when the price went down to five cents a pound. Even at ten cents a pound, a bale of cotton (standard weight of five hundred pounds) would bring only fifty dollars, and half of that would have to be paid to the landowner, leaving only twenty-five dollars for the family. If we made ten bales during the year, which was common, our annual income would be $250. To my knowledge, we never failed to pay off the debt that was incurred during the spring of the year.

My dad was also the community barber, using hand clippers and very primitive instruments. But the additional income helped immensely. At fifteen cents a head, he usually gave twenty haircuts each month, which totaled three dollars. Now, that doesn't sound like much, but back then, three dollars could buy quite a lot.

Our landowners would allow us to go into the woods and cut crossties during the winter months. We would cut down a hardwood tree, and then my dad would hue the crosstie to proper dimensions and sell it for a dollar.

Particularly at harvest time, my dad was in great demand to help his neighbors with their crops. The standard wage was one dollar for a ten-hour day. When I became old enough to lend a hand, I was paid fifty cents an hour. Sometimes, Dad would help a neighbor kill a cow or hog, and by helping, he would get a portion of the dressed meat.

Back in those days, people really helped each other. Our neighbors and landowners, the DuBoses and the Gambrells, were not much better off than we were, yet they often gave us extra groceries and food items that they had.

Also, we had a few chickens running around the yard, and a hog or two. Thus, we had eggs and fatback everyday, and occasionally fried chicken or a piece of ham.

My dad and I were avid hunters; not only for the sport but also for the meat it produced. Rabbits and squirrels were plentiful, and they made very good eating. We very rarely had quail, since shotgun shells were expensive and we never used them unless we were virtually sure we would get our game.

I did most of my hunting with a single shot 22 rifle. Shooting squirrels required accurate aim, and I even remember one time bagging a rabbit as it jumped up from its bed and ran away.

Our neighbor, Carl Lee, had several beehives and, as he harvested the honey, he would bring us a frame or two. The honey was a real treat.

Most of the time when we lived at both the DuBose and Gambrell places, we had no cow. Mrs. DuBose, Cantey's mother, separated the cream from her cow's milk and sold us the skim milk for five cents a quart. I did not care for skim milk, but it was better than nothing. Now, I'm told that it is much healthier for a person to drink than whole milk.

Also for five cents a quart, Mrs. DuBose sold us kerosene, our primary source of light. We had no place to store any extra.

When we moved to the Gambrell place, we had a huge pear tree in the middle of a large level field and also a smaller one nearby. When the pears were ripe, my dad would get out the wagon, my cousin Bill and I would gather the pears, and we'd take a wagonload into Walhalla, primarily to peddle door to door in the mill village.

"You wanna buy no pears?" Bill always asked, knocking on a door. We made money selling the pears. And during watermelon and cantaloupe season, we did the same thing, selling watermelons for ten cents, with the extra large ones going for as much as twenty-five cents.

The little house we lived in at the Gambrells was built on a hillside and consisted of three small rooms. One end of the house was elevated about six feet off the ground. In rainy weather, we played under the house, and that was also the place for the dogs to sleep.

The kitchen was a step down from the other two rooms. We had a wood-burning stove, a table where three or four people could eat, and a washstand on which a water bucket and wash pan always stayed. With no running water, we carried water from the Gambrell's well, which was about three hundred yards away. We heated the water for our Saturday night baths on the kitchen stove.

Our living room consisted of a fireplace, with firewood stacked in the corner, a table with several straight chairs, and a bed. Mother's sisters, Millie or Zeiba, and sometimes her brother Champ often used the extra bed.

The bedroom contained two beds and a trunk where we kept our most valuable possessions, if any.

Except for the stove in the kitchen, the house was heated entirely by the fireplace. As winter approached, we cut down trees in the woods and "snaked" them into the backyard, where we used sawhorses to cut the wood to the proper length. We used hardwood for the fireplace, and in the kitchen stove, we burned pine that had been cured for a year or so.

Pinewood ignited easily and quickly provided heat, which is important when cooking a meal in a hurry. In the long winter evenings, we popped corn, using a handheld wire mesh popcorn popper over hot coals in the fireplace. We also had black walnuts and sometimes chestnuts that we cracked on the hearth.

Some fifty yards or so behind the house, we had an outhouse. It was a cold place to go in winter.

This was very primitive living. However, everyone in the community was in pretty much the same condition.

A few of the landowners had better houses, and one or two even had a car. But no one had electricity, running water, indoor bathrooms, central heat, air conditioning, washing machines, or any of the conveniences that we take for granted today. Until 1936, we did not even have a radio.

On Sundays, we would often go to my mother's folks, the Whites. This was a special treat for me because I would have someone to play with or listen to. When it was time to go back home, Millie, my mother's sister, would fill up a big basket of food, which we found useful throughout the following week.

Unfortunately, we rarely went to church on Sundays. Mother said that our clothes were not good enough. But when I became a teenager, I would go with Earl Addis and his family on Sunday evenings to the Coneross Baptist Church, and sometimes with the Lee family to Poplar Springs.

Uncle Bub and his wife Dess had six children, and it was difficult to provide for such a crowd. So for that and other reasons, my cousin Harold began staying with us during the latter years.

Harold and I were addicted to going to the picture show every Saturday afternoon. Occasionally we would work on Saturday morning, but we always had Saturday afternoons off. There was always a Western movie playing on Saturdays. Buck Jones was one of our favorite stars, and our favorite serial was *Tail Spin Tommy*, the exploits of a young pilot, which ran for twelve episodes.

On one particular Saturday, nobody in the house had ten cents. Harold and I were heartbroken, until Dad suggested that we try to borrow the ten cents from Cantey DuBose. Sure enough, Cantey made us the loan, so we didn't miss a single episode of *Tail Spin Tommy.*

Back in those days, we thought nothing of walking from Bear Swamp to Walhalla, and occasionally to Seneca. But usually, we would catch a ride with someone. I remember one time when my mother, father, Harold, and I walked into Walhalla to see a Shirley Temple movie, and returned home late at night.

Another time, I had managed to save five dollars through working for others and selling blackberries, pears, watermelons, and rabbits. I bought an old bicycle from my cousin Bill, who lived in the Walhalla Cotton Mill village. Harold and I would ride this bicycle from Bear Swamp to Walhalla, and on up to Stump House Mountain, toward Highlands, N.C.

Going up the mountain was hard pedaling, so we would take turns riding. But coming down was a different matter. We would both ride, going at a high rate of speed and coasting almost the entire way back to Walhalla. This sounds dangerous today, but back then, there were few cars on the road.

While living in the Gambrell house from 1933-35, we bought our first cow, a young heifer that had horns barely sticking out of her head. We named the cow "Spikes" and the name stuck She was a Jersey cow and she gave very rich milk. Thanks to the DuBoses, whose land we did not live on, they let us keep our cow in their pasture. I can recall many late afternoons going all over Bear Swamp, trying to find Spikes.

During these Depression days, our breakfast would consist of hot biscuits with butter, an egg or two, fatback and gravy, and especially grits. I recall many times coming home from school about 3:30 or 4 P.M., and looking in the warming oven of the stove to find only left-over grits. Cold grits were not too bad when you were hungry. After eating the grits, I would go into the fields to pick cotton or do whatever work needed to be done.

In spring, summer, and fall, though, we had plenty

of fresh vegetables. Canning was done mostly during laying by time in August to provide for our main source of food in wintertime. At our house, unlike some others, we never got so low that we had only cornbread for breakfast. And our evening meal consisted of vegetables, whether fresh or canned, and some kind of meat. Sometimes it was wild game, ham, or chicken, or if nothing else, some fatback.

A special treat during these years was salmon mixed with eggs. Salmon cost five cents a can. Another delicacy was pork brains and eggs.

When we moved to the old Addis place, we went up a big notch in our Bear Swamp standard of living. Dad rented the property from Jess Addis, the youngest child who had grown up in the house. In the 1960s, the house became a popular restaurant.

Even though old, the house was in good shape. We had a nice well on the covered back porch and a nice outhouse just a short distance away. The front porch extended from one end of the house to the other. Mother and Dad had a bedroom, and I had a separate bedroom all to myself. There were other rooms, as well, with two fireplaces—one in the living room and another in the den. The kitchen was large, and there was even a room that served as a smokehouse, where we kept meat and other food.

During the second or third year we lived at the Addis place, my dad bought our first radio—battery oper ated, of course. The battery was a wet cell, similar to that in an automobile. It had to be changed quite often, but even then, we could pick up the signals from Greenville, Anderson, Atlanta, and Nashville as clear as a bell.

Upon moving to the Addis homestead, we bought, or had already bought, a horse named Maude, a wagon, and all of the farming implements and tools. For some reason, my dad always preferred horses to mules. He said they had more sense, and he was an expert at handling horses. We named our horse Maude for Carl Dubose's wife. They were both good-looking ladies.

I do not understand how my father managed to buy all of this, except that the people he bought from, either the Bauknights or Nevilles, extended him credit over a number of years. We only had four crops from 1936 to 1939, the year my dad's leg was broken and he had to quit farming. So apparently, the equipment was paid for during those four crop years.

When everything started so well at the Addis place, I was able to help Dad. We rented land from the

Mozers over on the east side of Richland Road, where we raised cotton and wheat. We also rented cotton land west of the old Burley place, where Mrs. Sertain lived. This was not a great distance from our home, especially when we walked through the woods and pasture.

One of the events that I especially remember happened on April 6,1936, when a terrible tornado wiped out Gainesville, Georgia, about sixty or seventy miles from where we lived. At our house, papers and all kinds of debris drifted down from the heavens. I was fourteen at the time and a sophomore at Walhalla High School.

About the only jobs available outside of farming were on road and bridge construction. Ten people would be lined up for every job. The old wagon roads connecting Walhalla to Highlands, Seneca, and Westminster were paved and other roads built.

When we went into town to shop, we left our mules and wagons out in the middle of the divided street, where there was a watering trough. As cars came into use, the watering trough and hitching post were moved behind the stores near the old livery stable, on the south side of town.

Baseball was the king of sports during this era. We had a Bear Swamp team that played in the cow pasture at the old Abbott place. We did not have a high school team. However, I did play right field on an American Legion team. A mill in Walhalla always had a baseball team, and many Saturday afternoons during the summer, my dad and I would take in the game. We would go also to Lonsdale at Seneca or to Oconee Mill in Westminster, when Walhalla was playing out of town.

During these Depression years, young people continued to marry. They would live with one set of parents until they acquired the minimum of household furnishings, which included a stove, table and chairs, one bed, one trunk, and a few pieces of silverware. If they did not own a house, which was usually the case, they could always find an old tenant house.

There were few church weddings in Bear Swamp during the Depression. Those were reserved for the wealthier families. A couple would simply go to a Justice of the Peace or to the preacher's house and, with one witness, the two would be married and announce it the next day to the family.

While we were living it, times were pretty hard back in the Great Depression.

"Buddy" with Uncle Alonzo White, his father, Berry Stegall, and mother, Weedie White Stegall

Cousin Bobby with Buddy Stegall

Chapter 6

Bear Swamp Folks

Bear Swamp is a small creek that starts as an even smaller branch on the south side of Walhalla, adjacent to the old Victor Monaghan Cotton Mill. By the time the water gets down to Bear Swamp Road at the DuBose place, the creek is fairly large—big enough to fish in and deep enough to swim in during high water.

When I was a child, the creek had an unusually wide delta, considering the actual size of the creek. While we lived at the old Addis house, I had a favorite dog, a blue tick hound named Jeff. He became sick, and no amount of doctoring could save him. I buried him down in the Swamp Creek delta, the spot I always visit when going home.

During the WPA days of the 1930s, the swamp was drained and channels were built. Prior to the WPA project, a sewer system had been put in place, to serve the south side of Walhalla, and especially the cotton mill village. Raw sewage was dumped into Bear Swamp Creek without the benefit of its going through a sewerage treatment plant. It is hard to believe how so many of us survived, living nearby and playing in the creek. But perhaps it was because we had our series of shots each year, especially typhoid.

Today, the area bears little resemblance to the old days, when it was all just one big swamp. The former farmland is now covered with forests, and there is little runoff from crops or sewage. Bear Swamp Creek has turned into a small, innocent looking branch.

Even after the Blue Ridge School was built in the early 1900s, the area surrounding it continued to be called Bear Swamp. To the families who lived there, it was more than a geographical area. It had its residency in our hearts as the place where kindred people lived and would always remember as home. Some families were outgoing; some kept to themselves. A few of the men drank too much, and some of the children did not turn out as well as they should. But the vast majority worked hard and provided the best they could manage for their families. We accepted each other, and we all pitched in when someone in the Bear Swamp community needed help.

My father and my family had a lot of friends back in those days of the 1920s and '30s—friends that I consider to be Bear Swamp folks. Here are a few of them.

DuBose

The DuBose family was a large one—twelve children, who were ruled over by Mrs. DuBose, a widow for many years. The best friend that our family probably ever had was Cantey DuBose and his wife, Felten. Their son, Ted, and I have kept in close touch, even to this day.

Back before my time, Monroe Gambrell had sold twenty-five acres of land to the DuBose family. This was located along the south side of Bear Swamp Road, where David and Richard DuBose now live. Before Cantey DuBose passed away, he divided this acreage between his two boys.

Gambrell

Monroe Gambrell and his wife Janie were our closest neighbors most of the time. They had a very nice farm with two tenant houses. We lived in one of them, and for a few years, my dad's brother, Uncle Bub, lived in the other.

In 1927, Monroe Gambrell and my dad were going to West Union in a two-horse wagon, when they were hit from the rear by an automobile. My dad was hurt very little, but Mr. Gambrell's injuries seemed to linger. He passed away the next year. Mrs. Gambrell and her unmarried daughter Carrie lived there, and Carrie became the business manager for the rental properties.

The Gambrells had four children near the age of my own folks. One of the DuBose boys married a Gambrell. Another son, Blake, lived with his family in Birmingham.

Paul Gambrell was the nephew of Monroe. He and his wife, Esther (whose maiden name was Arve) and their children lived up Bear Swamp Road, up from the Kiels, and adjacent to the Burley place. Paul ran a little store at his home.

Next door to the Gambrell farm stood an old farm that we called the Vernor place. I believe the Gambrells bought it in 1931. A black man, Lyce Davis, had owned and farmed the land for a number of years, before he sold it to Vernor in 1924. On the hilltop was an abandoned old house with trees and vines growing around the building. Also, there was a deep well that all of us boys had been warned about, to make sure that we did not fall in. While we were at the Gambrell's, no one farmed this land, so it was an excellent place to rabbit hunt and shoot quail.

On the original Gambrell place, there was once a tenant house on the south side, near the big pear tree. I believe the black tenant who lived there was named Pierce Cleveland. His wife had died, so on the way to town one day, he stopped to ask a neighbor's boy to go and baby sit with his three children, whom he had locked inside. Before the boy arrived, the house caught fire, and all three children perished.

For years, this was the tragedy most talked about in Bear Swamp. This old house was never rebuilt. Instead, another was erected a short distance away.

Another story of the old days was related to me several years ago by Robert Gambrell, the youngest son.

He and his brother John, together with his mother, were traveling in a two-horse wagon from their house to the general store, located at Coneross. I remember this store, which was across the highway from Coneross Church. It was operated by Clint Barker, who later moved the store about three hundred yards closer to his house.

Robert, John, and Mrs. Gambrell had loaded their wagon with eggs, butter, and other farm produce to barter for staple groceries, which they could not raise on their farm. As they traveled south, they had to go down a rather steep hill for quite a long way, and then make an abrupt turn at the bottom of the hill, so that they would not run into Bear Swamp.

The mules started traveling a little too fast, and as they got to the intersection at the bottom of the hill, the wagon overturned, breaking all the eggs and playing havoc with the rest of the produce. Fortunately, no one was hurt. The road, later changed to avoid this dangerous intersection, is now completely abandoned.

Lee

The old man, Fate Lee, had several children near the age of my parents and the one we knew best was Carl. Others were Jim, Zet, Bill, and "Punk." Jim married Cantey DuBose's sister-in-law. Bill married Ruby Addis, daughter of Warren. "Punk" married Rachel Addis, daughter of Floyd.

Carl and his wife, Ruby (Snead) lived on the Richland Road at its junction with Bear Swamp Road.

They had four sons, Morris, Frank, Broadus, and Boyd. Their daughter Virginia was only nine years old when her mother Ruby passed away in 1935. Virginia did yeoman's work to look after the family. Carl later married a fine lady, a schoolteacher known as Miz Bessie. Upon her death, Miz Bessie left the farm, which she had inherited from Carl, to her own relatives, rather than to any of the children.

Addis

At Blue Ridge School, there were more students named *Addis* than any other surname.

Earl Addis and his wife Maggie lived across Bear Swamp Creek from us when we lived at the Gambrell place. They had ten or more children, some of which were friends and playmates of mine. Maggie died as she went with her youngest child, Jack, to his first day of school at Blue Ridge.

Jess, Earl's brother, and his wife Minnie lived within sight of Blue Ridge School at the old Addis homestead. Their children were Gertrude, Jess, Jr., Dawson, and Butler. Grandmother Addis (Jess's mother) lived with this family for years, until she passed away at an elderly age.

Back in the 1920s, I had to walk a mile and a half over dirt roads and trails to get to school, and as I passed by the old Addis home place, I would see Gertrude and her three brothers. But in the 1930s, the family moved into town, and from then on, I did not have much contact with them, except for Jess, Jr., who was in my class.

Warren, another brother, lived in the house closest to Blue Ridge School. He had quite a few children. One was Lloyd, who finished Clemson shortly after I did.

Arthur, a third brother, moved his large family into the old Addis home place after Jess moved to Walhalla. I remember that one of the teenaged sons lost an eye when he exploded a firecracker in a bottle.

One of the Addis cousins, Floyd, lived for a while at the old Addis place immediately prior to our moving there. He had several children. The one I remember best was the oldest, Rachel, who married "Punk" Lee.

Yet another cousin, Miles Addis, lived near Poplar Springs Church. He had a large family, mostly boys. Miles won a new Ford automobile in a raffle put on by C.G. Jaynes Department Store. The sons nearest my age were Louis and the twins, Ennis and Ernest.

It is often strange how paths cross after so many

years. Gertrude Addis, the daughter of Jess, finished high school a couple of years before I did and went away to college. I did not see her again until forty-five years had elapsed.

She received her first college degree from Tift College, a master's degree from Furman University, a theological degree from Baptist Seminary in New Orleans, and a Ph.D. degree from Vanderbilt University.

Through those years, I lost track of her as she went into full-time Christian service. But my mother once mentioned that she had married Ed Tharpe, who had been a football star at the University of Georgia. He also had committed to full-time Christian work. The two served as missionaries in several locations in Hawaii, including the island of Maui, and at Kaneohe, Oahu, before going on to Hong Kong. Gertrude later explained that Dr. and Mrs. E.L. Morgan, former missionaries to China, had been very influential in her decision to become a missionary.

By 1980, Gertrude was living, temporarily, in Birmingham in a missionary home provided by Mountain Brook Baptist Church. She had a son who was in medical school in Birmingham, and also a daughter who was in the hospital after being injured in an automobile accident.

I called her and suggested that we get together. The Southern Baptist Convention was beginning its Evangelistic Convention that next Sunday night, with Stephen Olford, an outstanding and well-known Baptist preacher from New England and New York as the keynote speaker. Ann and I picked up Gertrude and arrived downtown at the First Baptist Church a little later than we should have. The only seats left were up near the front, on the second row.

Ann and I had known Stephen Olford for a number of years. He had often stayed at our house during the Keswick conventions. So it was a happy occasion to see him again.

As the service was winding down, the First Baptist Church minister announced that they were having a reception in the adjoining building for the convention officers and other dignitaries. He looked directly at me and said how pleased he was to have Dr. Tharpe in the congregation, and that he would like for Dr. Tharpe to get up at that time and join the others in the receiving line.

No one had told me that Gertrude had received a doctorate, and I assumed that, although I had met this Baptist minister on a few occasions, he had mistaken me for Dr. Tharpe. While I was deciding just what I should do, Gertrude got up and left to join the other dignitaries in the receiving line. I then realized how important and highly respected she was within the Southern Baptist Convention.

Gertrude has continued to live in Birmingham ever since. Her son, Dr. David Tharpe, is our neighbor at Shoal Creek. Her daughter, Jeannie, and her husband Glenn Shankles, also live in Birmingham. Her oldest son, Alan, is Dean of Reinhardt College in Georgia.

Although retired, Gertrude continues with a very needed ministry, providing grief counseling for those who have lost loved ones. Her husband, Ed, is deceased.

Gertrude and I talk quite often and, between the two of us, we have an understanding that her name is Gertrude and my name is Buddy.

How the world has turned and grown smaller from the time Gertrude and her brothers and I picked cotton in Bear Swamp.

BRANDT

John Brandt, who was from an old German family, married a Kiel, also of German origin. The family lived across the creek from the old Addis place. John ran a rather successful farm and dairy at one time. His daughter, Sophie, was about my age. His son, Rudolph, was killed in World War II, on board a ship in the Pacific.

KIEL

The Kiel family was one of the original German families in the Bear Swamp area. I particularly remember Maggie, who married Earl Addis. Old Mrs. Kiel lived in the family home place all alone. She always ran things with a firm German hand.

DEAL

The Deals, who came from Rabun County, Georgia, later moved into the old Brandt home, with their children.

BARKER

Garv Barker married one of the DuBose girls, Nell, and they lived along Bear Swamp just south of Blue Ridge School. Garv ran the syrup mill. Later, he and Nell moved into the old Addis place—after my father had broken his leg and we moved away. There were two boys in the family, Lawrence, who married Helen Barrett, and Edwin.

KIRBY

French and Sally Kirby lived at the old Abbott

place prior to the Moores' living there. Kirby and my father were cousins, with a mutual grandfather. They were also friends. Kirby was a good farmer and hunter.

GALBRETH

The Charlie Galbreths lived on Richland Road, except for the later years when they moved closer to Blue Ridge School. They had several children about my age. I knew their son, Matt, quite well. As young teenagers, we spent many happy days rabbit hunting and fishing down on Bear Swamp.

MEREDITH

Marlin Meredith and his family, including three children, lived just down the road from Arthur Addis. One son, Ralph Meredith, was about my age, but after the 7th grade, he dropped out of school.

WRIGHT

Charlie Wright was a special black friend of my dad's. They hunted and worked together often. Charlie lived on the backside of the old Abbott place before he moved into town, where he did well. He later had several children, who also made successes of their lives.

HOOVER

Another black family, who were friends of ours,

were the Hoovers. They lived in a tenant house of the Gambrell's. They had several children—two boys, Amos and George, and a girl about my age.

George and Amos were two of my best friends. We played together, worked together, and went fishing together in the creek.

Hoover and my dad got along splendidly. They would swap work with each other and between the two of them, they did a good job of farming the Gambrell place.

I recall one time when one of our neighbors had a misunderstanding with my father. He also had two very large and strong boys. I remember Hoover telling my dad not to worry, that he would stand by him if there was ever any trouble. With government aid, Hoover was able to build a barn and buy a mule and equipment, so that he could better his life.

WILLIAMS

The Williamses lived at the old Ansel place, just north of Earl Addis. They were renters, but also very good farmers. Their daughter, Nell, married Nathan Addis, one of Earl's sons.

SMITH

Earnest Smith lived with his wife and his wife's

sister, Cantie Kiel. The land that he farmed was not the best for crops. And he rather kept to himself. We lived on his place the year of 1926, when I was almost five.

Mozer

The Mozers lived in a large antebellum-looking house with columns. It was on Richland Road, close to Poplar Springs Church. The old man was a good farmer, and he raised several children, including sons Wash and Ike, who lived on the farm with him.

Cox

Mrs. Cox, a widow, lived in a house on the Westminster road, near Coneross Church. Her several children went to school at Blue Ridge.

Pepper

Also living on the Westminster road were the Pepper family. But they were not farmers. They had one son, W.P., who was in my grade at Blue Ridge.

Foster

Charlie Foster was my mother Weedie's uncle, although they were about the same age. They lived on the Westminster road also, with their several children going to Blue Ridge School. Their daughter, Sophie, was my classmate.

SANDERS~GRAHAM

Granny White's brother, Edward Foster, had a daughter named Maggie, who married Lawrence Sanders. Sanders died early in life, when the children, Julius, Walt, Gussie, Nacoda, and Nell were still very young. Maggie then married a Mr. Graham and they had several children, including Leonard, my third cousin.

Leonard went into the army near the end of World War II and stayed in for a time after the war. He was stationed somewhere out in the northwestern part of the country, when he was called back to Bear Swamp because of his mother's severe illness.

Leonard rode the train day and night for several days without any sleep. He finally arrived in Seneca, where his mother was hospitalized. When he reached his mother's bedside, she said, "Well, Son, I waited until you got here." Then, she passed away.

A short time later, an exhausted and upset Leonard went home. A few hours later, he also passed away. Maggie and her son Leonard had a joint funeral.

ALLEN

The Allens lived on the Seneca Highway, at the old Darby place. Although this was not considered a part of

Bear Swamp, their children went to Blue Ridge. I particularly remember their sons, Olen and Arthur, and their daughter, Rubye. The Allens had moved from the Salem community.

Kerr

I have lost track of this family and their children since World War II. The Kerrs lived at the junction of Richland Road and the Seneca Highway, near the Allens.

Burley

Fritz Burley and his sister, Maude, lived very near the Earl Addis place, just a short distance across the road from the old Bear Swamp School site. We learned to swim in Burley's Pond, located in the pasture. Mr. Burley's brother was our rural mail carrier through all those years.

Fricks

The Otto Frickses lived up Bear Swamp Creek toward Walhalla, not far from the Ansel place. Mrs. Fricks was a very pretty lady, with two very pretty daughters, Eleanor and Frances.

Moore

The Moore family moved into the old Burley place, where Uncle Will Stegall had lived for years. They had moved from a larger city—Greenville, I believe—to weather the Great Depression. They were extremely fine folks who had hit upon hard times. They had two boys about my age, Bill and George. But by the time I started high school, and times were better, I lost contact with them, since they moved back to the city.

Hall

The Halls lived on Richland Road, between the Mozer plantation and Poplar Springs Church. Their only child, Eliza, married Ike Mozer, and they lived in the antebellum Mozer house. Before she married Ike, Eliza dated Carl Lee, after he had lost his first wife, Ruby. She would have made a wonderful mother for Carl's young children, although at the time, she was little older than his oldest.

Blackwell

Virgil Blackwell and his wife had a son, Arnold, and also a daughter. The family lived up the creek from the Kiel place. The children went to Blue Ridge School, but they were much younger than I.

Logan

Henry Logan, his wife, son, and daughter lived between the Burleys and the Otto Fríckses. The children, Thomas and Pauline, went to Blue Ridge School. Henry was first married to a Foster, a relative to Granny White.

White

My grandfather, Wesley White, moved from their old home place, which they owned on Negro Fork Creek, to the Abbott place in the Bear Swamp community in 1923, in order to have more land to farm. By that time, my mother was already married to my father. Her brothers and sisters who lived at home were Alonzo, George, Thode, Zeiba, Millie, and Champ.

I was very close to Granny White and the rest of the White family. I will have more to say about them in a later chapter.

The Black Community

The Bear Swamp families would not be complete without mentioning the black community. I have already mentioned two of our good friends and their families—the Hoovers and the Wrights.

In an interview with Jess Addis in 1987, he spoke of the huge tract of land running approximately from Cane Creek westward to Bear Swamp Creek. After the Civil War, it was set aside by the government for black families. They were given forty acres and a mule. There is evidence to this day of some of these old home places.

When I was going to Blue Ridge School, Double

Springs Church, a very active church, was still in existence, along with a one-room schoolhouse. While we lived at the Addis place, the church and school were our closest neighbors. The graveyard for the church was adjacent to the fields we cultivated.

As to how the church got its name, it might have been because of a spring on each side of the ridge where the church was located. The church is said to have been founded by the family of Pierce Cleveland, whose three children died in the house fire.

Some of the black families later moved into town. And although he was not strictly a member of the Bear Swamp community, one of our favorite black friends was Peter Major. He had been a slave, possibly in the Stegall family, and he lived by himself on the edge of West Union.

West Union had stores, a cotton gin, and mercantile establishments just like Walhalla, so it was not necessary to go all the way into Walhalla to trade and take our cotton.

On our way, my dad and I would always stop and visit with Peter Major. He lived by himself, even at that advanced age.

There was always a feeling of mutual respect

between the white and black communities in Bear Swamp. We helped out each other and lived together in peace and harmony.

These then are the Bear Swamp folks that I remember. ☒

Berry with his plow team

Chapter 7

Uncle Bub, the Moonshiner

William Spencer "Bub" Stegall was the oldest son of Pap and Sue Stegall. Many have said that he was a very handsome man, a good dresser, and very popular. After Pap got too old to manage, Uncle Bub headed up the Stegall clan.

He married Bessie Chapman, apparently a very nice young woman, and from that marriage came Mae (who was slightly older than I was), Harold, Julian, who died at six months, Louise, Jack, Bobby, and Gene.

Bub had a hard time providing for his family, especially when the children were small. The family lived for some time in another tenant house on the Gambrell property. It was a very primitive cabin, built mostly of logs, and it had only two rooms. They did, however, have a well, and did not have to carry water a long distance as we did.

Bub and Dess's fourth child, Julian, was born at the cabin in the winter. But when summer came, and with no screens in the windows, the heat became an enemy, and sanitary conditions were hard to maintain. The young boy became sick and lingered for days. Finally, at six months, he died.

Since there was no money for undertakers or a hearse, Cantey DuBose used his car to take the casket from the house over to Coneross Church, where Julian was buried alongside his Grandmother Stegall and Aunt Ada. Several years ago, it was a pleasure to provide suitable markers for these three graves.

The very bottom of the Depression years, as far as I am concerned, was the time that my dad and I got corn out of the corncrib, shelled it, and took it to White's Mill, to be ground. We then took a sack of cornmeal to Uncle Bub's house, where there was almost nothing for them to eat. For me, the epitome of poverty was eating cornbread, and nothing else, for breakfast.

Uncle Bub was a tenant farmer, much like my dad.

He occasionally worked at Public Works projects. In later years, was a much-sought-after well digger.

He and his helpers, often his own boys, would dig a well big enough in diameter to actually work down inside. They would draw up big tubs full of earth, until they got deep enough to strike water. Since his work was excellent, he was always in demand.

Another successful operation, the cash crop of the Appalachians at the time, was his whiskey operation. He would have been even more successful at it, if he had not sampled his merchandise so liberally.

He would walk about ten to twelve miles from his home near Walhalla to the mountains, where he purchased whiskey from a person who ran a distillery. Returning with as many one-gallon cans of whiskey as he could carry on his back, Bub then stopped in the woods near his home and proceeded to empty the contents of the gallon cans into pint bottles or fruit jars. He would then hide the jars in stump holes, other holes in the ground, under logs, and even under rocks, all the time covering the jars or bottles with leaves.

On one occasion, when he was placing a pint of whiskey behind a rock, he was bitten by a large rattlesnake, which had already chosen the rock as a hiding place. The snake was poisonous enough to kill a man in a few minutes. But Bub turned up the whiskey container and drank it all. He survived the snakebite. However erroneous, the whiskey was later attributed to saving his life. This was a favorite family story for a number of years.

Years later, after his children had married and moved away, I went by to see Bub and Bess on a trip back to Birmingham. Bub was in amazingly good shape, although very aged. His moonshine days were over. He did not drink anymore, and he seemed to have a good standard of living. But, as a dedicated Christian, I was more concerned with his spiritual well-being. I engaged him in conversation about God, but when I left, I felt that I had failed to reach him.

A few more years went by. Then, I received a call that Bub had died and that the funeral would be held the next afternoon at four o'clock, at Pleasant Hill Baptist Church on Coffee Road—the old road that connected Walhalla with Westminster.

I was very busy in business at the time, and my wife Ann questioned whether I should attempt to drive all the way from Birmingham to Walhalla. But something strong, like a magnet, was pulling me in that direction.

On the day of the funeral, I went by my office for a while, and then got into my automobile. Without stopping on the way to visit friends or family, I traveled alone through Georgia, and into South Carolina. For miles, the Blue Ridge Mountains were visible to my left. What a beautiful sight! It was October, with the red and golden leaves at their peak. Surrounded by such beauty, I thought about Bub and his wasted life. He had once been full of dreams and ambitions, and they had been squandered. Some members of the family had said that if he had married the right person to encourage him, he would have become a success. But of course, there are always two sides to a story.

As I arrived at the church and parked in the area among the trees surrounding the church, the first person I saw was Mae, Bub's daughter, who had been one of my first playmates when we were children. She came running over and threw her ninety pounds around my neck.

"Oh, Bud, do you know where Daddy is?" she asked, calling me by my childhood name.

Her excitement, as well as her question, puzzled

me. I was thinking, "Mae, I believe if you went into the church and looked in the casket, you would find him." But fortunately, I did not say that aloud.

She repeated,"Bud, do you know where Daddy is?"

"Tell me."

"Daddy's in heaven," she said.

"How do you know, Mae?"

She told me that her preacher had developed a good relationship with her father. When he explained the Gospel to him, Bub had enthusiastically accepted the Lord.

So, what had started out to be a very sad day turned out to be one of the most glorious days of my life; for I realized that, while we are living, it is never too late to hear the Good News.

Uncle Bub

Chapter 8

Another Branch of the Family

In the Rock Springs community and along the Anderson-Greenville county lines, there is a saying that, if you are not a Stegall, then you must be related to one.

That area of excellent farming land was considered some of the best in South Carolina, prior to World War II. After the war, a lot of industry came in, thus phasing out most of the farming.

My dad had very little communication with his older half-brothers and sisters in the Easley area. But I did hear him speak of Martha, Lou, Boy, Jim, and his youngest half-sister, Mary Elizabeth, who lived most of her life in Oklahoma.

When I was about five or six, I recall going with

my parents by train to visit some of the Stegalls near Easley. We stayed in a very nice two-story house, across the highway from the Rock Springs Church. I can remember having cheese for breakfast, which was unusual for us. That was the only trip I made to Easley to visit relatives before World War II.

After Ann and I had been in Birmingham for several years, we learned of the Stegall reunion that was held each August at the Rock Springs Baptist Church. We went back several times, taking some of our children with us.

It was at the Stegall reunions in the late 1960s and early 1970s that we got to know my first cousin, Mary Hamilton Turner, and the four Stegall sisters who live in the Williamston area.

Mary was married to Cliff Turner, a barber. Our spirits meshed together beautifully, and I enjoyed every minute spent with Mary. She had a high regard for our grandfather, Pap Stegall.

Mary and Cliff lived in a house on West Main Street in Easley, which had formerly been one of the cotton mill houses. They always had an excellent garden, and it was quite a feast, having dinner at their home, which we did at every opportunity.

Mary told me many stories about the Stegalls, and I wish that I had made a tape, or at the least, kept better notes. She was the one who told me the story concerning Pap's cotton gin and the missing money.

A schoolteacher all her life, she also told me many interesting stories of going back into the mountains of Pickens County to teach in one-room schools. She lived with some of her students' parents, who had no electricity, no running water, no indoor plumbing, and no automobiles. Later, she taught at Pierceton, not too far from where the four Stegall sisters, Ruby Couch, Bessie Wyatt, Edna Stegall (who also married a Stegall), and Effie Wilson live.

The Stegall sisters are descendants of Pad Stegall, a brother to my grandfather, Pap. Patrick, Pad's son, was their father. So this makes us cousins, two or three times removed. The four finished high school, got teaching certificates, and continued teaching while getting degrees at Clemson University.

They are all well thought of in their community and are very active in their Baptist church.

Mary Turner was extremely fond of her cousin, Ezekiel Stegall, whom she called "Zeke." He held the very responsible position of judge in the Tax Court of the United States in Washington.

Zeke was the son of Thomas Stegall, a brother to my grandfather, Pap, and his wife, Ella Long. That would have made him a first cousin to my father, Berry, and to Mary's mother. Zeke had inherited the old home place of his father, Thomas, in the Rock Springs community.

Apparently, Zeke had quite a good education, with a law degree. He became well known in Washington, and married a lady who rather frowned on Zeke's rural family background. According to Mary, she attempted to keep Zeke away from the entire Stegall clan in South Carolina. When he became disabled and went to a nursing home, Mary was not allowed to talk with him, and her messages were not relayed to him. It was only through Zeke's pastor, a Dr. Patrick, of the Wisconsin Avenue Baptist Church in Washington, that Mary's messages reached Zeke.

He spent his last days in the Baptist home in Washington. Mary so badly wanted me to get together with Zeke, and I wish now that I had made an effort to do so.

Mary and Cliff Turner had two sons—Bill, who worked for Duke Power Company, and Thomas, with the Philip Morris Corporation in Concord, North Carolina.

After her husband Cliff passed away, Mary continued to live in Easley and to teach until a few years before her death. As her health began to fail, she moved into a retirement home in Charlotte, N.C., where one of her sons lived nearby. In January of 1987, when I last saw her, Mary was eighty-three years old.

The Whites, summer of 1922
(L. to R.) Champ, Granny Tina White (holding Buddy), Millie with Cousin Bill, Grandfather Wesley White, Weedie, Alonzo and George

Chapter 9

The Whites

My mother, the former Weedie White, was already married to my father when the rest of the White family, including her brothers and sisters, moved to the Abbott place in Bear Swamp in 1923.

Granddaddy Wesley White rented the Abbott property in order to have more land to cultivate with his four boys. But after Alonzo left home in 1926 to make his fortune in Florida, Granddaddy White decided to move back to his own home place a year later. He undoubtedly made a mistake by not buying the Abbott land, which had been offered him at payments equal to his rent.

Wesley White was a well-respected man in Oconee County. He was a member of the Masonic Lodge, which, at that time, was an exalted position in the county. He took his family regularly to services at Zion Methodist Church, and he worked extremely hard during the week.

His only hobby seemed to be reading. He kept up with news of the world by subscribing to *The Atlanta Constitution* newspaper, which was delivered by mail. I never knew of his going fishing or hunting or to ball games. When he was not working, he read.

My grandmother, Tina, although she had little formal education, was a wonderful, dedicated wife, mother, and grandmother. She had a world of home remedies that dated back to her part-Cherokee heritage. She also kept a 30-calibre Winchester rifle, with which to shoot hawks and other *varmints* that bothered her chicks.

During the troubled times at the White home place during the 1930s, both my aunts, Millie and Zeiba, stayed with us for a while, at separate times. We then lived in the DuBose tenant house up the hill from the Gambrells.

I recall that, during that period, Zeiba took a business course in Seneca and met a nice young man, who was a student at Clemson. He came often to visit her, and I have often wondered what happened to him. Zeiba later married Fred Newell, and they settled down in

Springfield, Massachusetts, Fred's hometown. The two had no children.

Millie was a second mother to me. From the day I was born, when she was ten years old, she looked after me many, many days while my mother worked in the field alongside my dad, picking cotton, or hoeing.

From the time I left my parents standing on the steps at the old Jess Addis house in Bear Swamp in November 1938, when I went away to junior college, coming home was never complete until I went to see Millie. She was always glad to see me, and she made me feel special. Millie simply loved me—as I loved her, and her coconut cake.

During the decade of the 1930s, the Whites had pretty much scattered. Champ went down to South Georgia in the fall of 1936. L.O. White, a cousin, had been living there for quite some time. He was a businessman who owned a taxi service, and possibly a farm, as well. He helped Champ find a job, but then when Champ developed pleurisy, he returned home in early 1937.

That same year, Uncle George went to South Georgia, where he made his home for the rest of his life. He began his career as a farm hand, working by the day on a farm near Cordele. Later, he married a farmer's daughter, Myrtice Lewis. They made an excellent match.

George's next step was to become a tenant farmer, probably farming on halves—half to the owner, half for himself. He progressed from that to owning his own mules and implements, and then rented land for an annual fee. Doing quite well at this, he saved enough money to buy his first little farm.

Several times he traded farms, each time for the better. He finally ended up on a 250-acre farm, east of Vienna, Georgia. As new machinery, such as cotton pickers, came on the market, George was one of the first to buy. The equipment enabled him to grow many more acres of cotton and peanuts with a minimum of labor.

George and Myrtice had three children, Wesley, Grady, and Linda. As the boys grew older, he bought adjoining land and rented other acreage. George and Myrtice sent all three of their children through college, some receiving advanced degrees. But Wesley, who was the only one who truly loved farming, elected to go through a two-year technical trade school in St. Louis, where he learned welding and other skills that would help him on a farm.

Wesley now has one of the outstanding farms in Dooley County, with irrigation systems, cotton pickers, and other systems. Last year, he cultivated five hundred acres of cotton and four hundred acres of peanuts. He calls this a small farm. This is true when compared to those of several thousand acres.

About thirty years ago, when he became partners with his father, and began to prepare an income tax return, Wesley discovered quite an unsettling fact. His father, George, had never filed a tax return, either federal or state. Growing up on a few acres of land in Bear Swamp before many of the tax laws were enacted, perhaps he was not even aware of the necessity for farmers to file a return. Wesley asked for my suggestions, and I had them come to Birmingham to meet with David, who was my accountant.

I had complete faith in David. He did a magnificent job in digging into all the years gone by, helping Uncle George at a very minimum cost.

Several months after David became Uncle George's accountant, the IRS came calling. The IRS intended to make an example of the old man. "You just simply cannot refuse to make tax returns." They told my accountant that they were going to send Mr. White up for a long sentence.

David quietly explained that the old man was one of the most respected farmers in Dooley County, and that there had never been any question about his honesty and dependability. He also warned that a court jury would consist of his peers, who loved and respected him. The IRS agreed to settle for a sum of approximately ten thousand dollars.

A few years after this event, Uncle George, who had been extremely active and agile all his life, was suddenly stricken with crippling arthritis. He had both knees operated on at the Medical Center in Augusta, but they never healed completely. He became an invalid, and finally moved to a nursing home.

Several months before he developed arthritis, an insurance salesman had sold Uncle George a disability policy that paid a considerable amount of money monthly.

A few years before George passed away, Bud and Elizabeth Newbold, and Ann and I went to visit him. I made a statement that I really would like to have some real milk, fresh out of a cow, before the flavor was cooked to death through pasteurization.

I was told that selling raw milk was against the law, but that the farm adjoining George's was a dairy farm. The owner would not sell me a gallon, but might consider giving me one. We went to see him, and sure enough, he graciously gave me a gallon, fresh from the cow. The Newbolds turned up their noses, thought I was an idiot, and would poison myself. We went on to the mountains, where I began to drink the refrigerated milk. Seeing my obvious enjoyment, they also wanted to try it. They ended up drinking most of my gallon of raw milk.

Perhaps visiting with Uncle George brought back childhood memories of my grandparents, Wesley and Tina White, and those days when I walked across the field from Blue Ridge School at Big Recess to eat lunch at my grandparents' house. Always on the table was a large pitcher of fresh milk.

Granny White

Chapter 10

Corbin~White Connection

W.B.F. Corbin became a U.S. Deputy Marshal in the late 1800s. His main concern was fighting the illegal alcohol, or moonshine, trade, especially back in the mountains. Born and raised in the Salem community near Walhalla, he had many friends. But because of the nature of his profession as a Federal Revenue officer, he also had quite a few enemies.

Early one day in 1901, as he was returning home, he was surprised by several moonshiners, who had decided to rescue a couple of mules that had been seized in a raid. The moonshiners opened fire, and Corbin was hit on the chin and his left arm with shotgun pellets.

Corbin pulled out his 45 revolver and returned

fire. He was wearing a heavy overcoat, so although other shots were fired at him, they did little damage. Corbin's sons, hearing the fight, brought more guns from the house, and the moonshiners were driven back into the woods.

By 1909, Marshal Corbin was still hunting down lawbreakers. A man named Bob Belcher was accused of mutilating a rural mailbox. Back in those days, to damage an RFD mailbox was a serious Federal offense. So Corbin, with his half-brother Chris, whom he had deputized, went in search of Belcher to arrest him.

On March 8, 1909, the two found him at the home of J.B. Palmer on the Westminster-Walhalla Road. In the process of arresting him, both Corbin and his half-brother were killed. Belcher was found guilty of the crime and was sentenced to life in prison. The sentence was later reduced to twenty years.

The story of Corbin's life was the subject of a book, *Peter Corbin and his Descendants,* written by William Clyde Whitten, Jr., of Walhalla/Clemson. The descendants lived in the family home on Coffee Road at least until the death of his daughter, Alice, who was my first grade teacher.

In an attempt to research the genealogy from my mother's side of the family, I ran into a possible connection with the Corbins. W.B.F. Corbin's mother was a Nicholson from the Salem area. Ted DuBose, the son of Cantey, married Brenda Nicholson from Salem. Brenda's mother, Estelle Moss, was a good friend of my mother, and they considered each other cousins. Estelle's mother was Minerva Ross, a sister of Jane Ross, who married my great-grandfather, Theodore Alexander White. Our family tree indicates that Brenda Nicholson DuBose and I have mutual great-great grandparents, John J. Ross, Sr., and Melissa Martin Hudson.

W.B.F. Corbin and his wife Mary Susan (Burgess) had a large family. Ten of their twelve children lived into adulthood—Jim, Elbert, Lilly Blake, Mary, Sarah, Cornelia, Henry, George (Doc), Alice, and Fred.

Jim Corbin moved to Birmingham, Alabama about 1905 and had a good job in the mines. Cornelia, his sister, also moved later and married a man named Joe Box. While we lived on Biltmore Avenue in Vestavia, our next door neighbor was Joe Box. But he could easily have been from another family not related to Cornelia. At the time, I was not aware of any possible connection to the Corbin family.

In the 1920s, two other Corbin brothers, Henry

and Fred, followed Jim and moved to Birmingham, where they worked as railroad shop machinists.

During the long winter evenings in 1925, the three brothers, Jim, Henry, and Fred, must have sat before a fire and reminisced about their lives in Walhalla. This was also the Depression era of bank robberies, particularly by the notorious Dillinger gang. The three began to discuss how easy it would be to rob the little Enterprise Bank back home in Walhalla. And from that conversation came the planning of the deed, and the change from a law-preserving family tradition to that of law-breaker.

The Corbin brothers either drove a car or rode the train back to Walhalla. Breaking into the bank was easily accomplished. They took all the money that they could carry, including the silver and even the pennies. The story goes that the three were tracked by the trail of coins they dropped along the way.

They were soon apprehended, stood trial, and were sentenced to the State Penitentiary. Fred received a lighter sentence, probably due to his age. He remained in South Carolina and from then on was a law-abiding citizen. When Jim was released, he returned to his family in Birmingham. Henry was released from prison in 1931.

Millie, my mother's younger sister, and Alice Corbin Elrod had been friends for years. It was through Alice that Millie and Henry got to know each other.

They were married in 1932. Henry had been married in Birmingham, but the marriage had ended in divorce. Millie had a two-year-old son, Bobby, and they lived with Millie's parents, who were my grandparents, Wesley and Tina White, along Negro Fork Creek. To this marriage, Gene was born on August 21, 1933.

Also living at the old home place were Millie's brothers and sister, George, Champ, and Zeiba.

Zeiba, at that time, was dating a man named Volrath. That was not a good choice, since he had been married and divorced and because of that was not highly thought of. South Carolina at that time did not allow a divorce or recognize one from another state.

On long winter evenings, George and Champ White and Henry Corbin would often go over to Jess Harvey's house to play a card game called "Setback." The Harvey house was built on the crest of a hill, and the men had to cross a branch to get to it. It was not unusual for men, traveling through the countryside at night, to take a weapon with them.

On this particular night, the three, armed with a 22 rifle, were returning home about 11 P.M. Nobody is quite sure how it happened, but apparently the three men were startled by Volrath's car, which almost ran over them, and they thought it appropriate to give Volrath a scare by firing at the car.

They went on home, but before retiring, they noticed the headlights of a car that had run off the road connecting the Westminster-Walhalla Road to the Negro Fork community.

They went to investigate, and discovered Volrath's car in the gully, and Volrath dead.

Henry Corbin admitted in court that he fired the one shot from the 22 rifle at the rear of the car as it moved down the driveway. The low-powered bullet had penetrated the steel, traveled through the back seat to the front seat, and then lodged in a vital organ in Volrath's body.

Granddaddy White mortgaged his 60-acre farm so he could hire the best lawyer in the county. All three men were arrested and put into the Oconee jail in Walhalla. Charlie Burnside, an old friend of George and my father, put up bail for George's release. Champ stayed in the jail a little longer.

I remember, as a twelve-year-old, going to the jail with my father to visit Champ. We were locked inside the jail with thc prisoners. Champ was finally released on bail, but Henry remained until the trial. He was sentenced to three years in the State Penitentiary. After serving half that time, he was paroled.

Paying for the lawyer put such a financial burden on my grandfather White, that he was never able to pay off the debt in his lifetime.

Henry and Millie returned to live at the Whites. On December 8, 1936, their daughter, Joyce, was born. During this time, Henry received his WWI veteran's pension. He bought an A-Model Ford, and went back to his old job as a machinist in Birmingham. Granny White was heartbroken when they took the children away. But Millie and Henry soon returned to South Carolina

When Zeiba completed her business course, she began work in Anderson. There, she met her future husband, Fred Newell. After many years in Anderson, they moved back to Fred's hometown in Massachusetts.

On Saturday, April 8, 1939, Granddaddy White was mending fences. Most country people took off on Saturday afternoons to go into town and visit with other farmers, buy a few goods, or else go to a ball game. Henry did the same.

No one knows what actually happened. It could have been that when Henry passed the fence row that was being mended, Granddaddy White might have mentioned that he could use a little help and perhaps criticized Henry for taking off. A fight ensued. Henry then proceeded into town and returned later in the day.

Far past the time that Granddaddy White should have come home, the family became worried and began to look for him. They found him by the fence with the ax, that he had been using, lodged in his skull. Death had been immediate.

Champ accused Henry, and the sheriff arrested him and took him to jail. By this time, Henry had acquired a rather bad reputation, and when word got around as to what had happened, the sheriff, fearing a lynch mob, took Henry to another county and another jail.

Most of the White family thought that the deed was cold-blooded, premeditated murder. Somehow, I feel that it was not planned, that Henry more than likely let his temper get the better of him, and he didn't realize the seriousness of the fight.

Nevertheless, Henry was convicted of first degree murder and sentenced to life in prison. He was never released on parole. The only White who ever visited him was Uncle Thode. Henry is buried in Westview Cemetery in Walhalla.

How different from his sister Alice, a fine woman and teacher, with outstanding children, who have followed the earlier family tradition of service for the betterment of the community.

Wesley and Tina White with Millie Corbin's children – Bobbie, Gene, and Joyce, 1937

Chapter 11

George Foster and Emily Walker

George Foster and Emily Walker were my great-grandparents, and the mother and father of Granny White. Both were born in the Spartanburg, S.C. area, and both were part Cherokee. My mother said that Emily was at least one-fourth Cherokee. My Uncle Thode believed that both were one-half Cherokee. Winford Foster, one of their grandsons, mentioned that Emily served as a doctor for the family, because of her knowledge of medicinal plants and herbs, used in old-fashioned Indian remedies.

George and Emily are buried at the Neville Cemetery across the Walhalla-Highlands Highway from the Rocky Knoll Baptist Church. Zion Methodist Church stands on the other side of the cemetery.

Sometime around 1870, George Foster served as a gauger working for the U.S. government at liquor distilleries. This was before prohibition, of course. I imagine his job was to inspect for cleanliness and make sure of the purity of the ingredients. In modern times, I'm sure we would call him an inspector.

There was a man working at this distillery who was evidently disgruntled about the management, plus being just plain mean. He sought to get even by putting a buckeye into a ten-gallon keg of whiskey. George saw the man's action.

Possessing a buckeye was supposed to bring good luck, but as children, we were always warned never to eat one, because a buckeye is extremely poisonous.

Foster reported the incident to the management, and so the man was fired. From that time on, the man sought to get even with George.

On his way home, George Foster always used a path that meandered through an undeveloped area. Since he had been forewarned that the man was out to get him, he was cautious. One afternoon, as he was walking home, he saw the man, who had been lying in wait to do him in. But while waiting, the man had evidently drunk too much and had gone to sleep, with a long, sharp knife lying across his stomach.

According to Uncle Thode, Foster simply picked up the knife, slit the man's throat, and then walked on.

No one formally accused Foster of killing the man. However, the authorities suggested that it would be best for Foster and his family to move out of town and take a similar job at a distillery in Walhalla. In those days, it was a long way from Spartanburg to Walhalla, requiring several days of travel by wagon. Foster immediately went to his new job at Walhalla, while the family followed several weeks later, with their possessions in a two-horse wagon.

I have gone to Cherokee, N.C. to look through the Indian archives for any information about Emily Walker. I found the name several times—once on the Siler roll made in 1851, and also on the Chapman roll, made in 1852. Could this have been my great-grandmother?

The Fosters originally came to the Spartanburg area from Virginia. In my file on the Fosters, I have an unusually good picture, made in 1940, of my grandmother, Tina, and her brothers and sisters—Pierce Foster, Dora Foster Harvey, Charlie Foster, and Lou Foster Rogers.

Great-Aunt Lou was Tina's youngest sister. When

she was quite young, she married an elderly Confederate veteran named Rogers. She had one son, Joe Rogers, who was an outstanding citizen in Walhalla and Oconee County for many years.

She was quite an eccentric character. Some thought she was a little touched in the head, but many of us believed that Great-Aunt Lou had a superior mind—and that it was her intelligence that set her apart.

Granny Tina Foster White with her brothers and sisters, Pierce, Dora, Charlie, and Lou Foster

Chapter 12

School Friends at Blue Ridge

In the early 1900s, three small grammar schools, Bear Swamp School, Coneross, and Poplar Springs, were consolidated into the Blue Ridge School District.

Two of those earlier schools had been built adjacent to a church—Coneross, across the Westminster-Walhalla Highway from Coneross Baptist Church; the Poplar Springs School adjacent to the Poplar Springs church on Richland Road.

Besides the Coneross school and church, there were also in that community a store and a few other businesses. Although the old school building is no longer there, the Coneross Baptist Church is still going strong, and the congregation has recently built a new sanctuary just south of the old one. Many of my White and Stegall relatives are buried there. The church in the Poplar Springs area is also still vital, with a new sanctuary.

But the old Bear Swamp School stood alone, with no church adjacent to it. Most of the earlier people who had settled in that community were of German origin, and they went to the Lutheran church in Walhalla. The old school's former site is located across the Walhalla-Bear Swamp Road, opposite the driveway to the Fritz Burley place, and about five hundred yards northwest of the new Highway 11.

The Blue Ridge School, drawing students from all three communities, Coneross, Poplar Springs, and Bear Swamp, at first provided education from the first grade through high school. As in many other farming communities, the school year consisted of a short summer session of six weeks, and then was recessed for cotton picking until late October, when we all returned

Some of the older DuBose and Gambrell children graduated from high school there. My first two or three years of grammar school were also spent in that building. But then, disaster struck.

One afternoon in the summer of 1930, most of the

farmers of the community had decided to go fishing after the hard work of laying by their crops. So when lightning struck the school that day, there was no one around to help put out the fire. The school burned to the ground.

The men got together, and a new school was rushed to completion for the fall term. But from that time on, Blue Ridge was used only as a grammar school for grades one through seven.

There were only four rooms rebuilt. Even then, no more than two were used as classrooms—one, for the first through third grades, and the second, for the fourth through seventh grades.

My first grade teacher was Mrs. Elrod, the former Alice Corbin. She was also my second grade teacher, and she stands out above most of my other teachers. She was a great encouragement to me in that little country school.

By the time I reached the seventh grade, there were only about eight or ten students in my class. As I recall, I was the only one from the class to go on to high school in Walhalla and to graduate.

Blue Ridge School closed in 1950. Buses now carry all the community schoolchildren into Walhalla.

I attended all seven grades at Blue Ridge. Since we

had no other transportation, I always walked to school. Oddly enough, I never missed a single day while in grammar school. And I received only one whipping, which I probably deserved.

Before I started to school, our closest neighbors with children lived a mile away. My only playmates at that time had been my cousins, Mae, Edna, Ruby, and Bill, which I saw only every few weeks.

Then, when school started, I suddenly had new friends and classmates. Earl Addis, who was in the same class, lived closer to me than any other schoolmate. We became good friends and spent many hours together—fishing, playing ball, stripping the fruit from muscadine vines, and anything else that would interest a young boy.

Earl was from a very large family that included his brother Nathan, Nathan's twin sister Eugenia, Margaret, Kenneth, Harmon, and Jack. I loved to spend time at the Addis house, which was located across Bear Swamp Creek from the DuBose home.

Earl did not finish high school. He joined the army at the beginning of World War II, and saw much combat with the Fourth Infantry in Europe. He was wounded several times and was discharged after spending years in the hospital. Finally, the effects of his war wounds caused an early death, some ten or fifteen years after the war.

Living in the other direction were Frank Lee and his brother, Morris. Both had polio as young children, and the disease left them both with disabilities. Although Morris, from that time on, had the mind of a five-year-old, he was a very dear, sweet person. Frank had a withered left arm, which was a real handicap when he was playing baseball or doing chores around the farm.

Also in the family were a sister, Virginia, and two other brothers, Broadus and Boyd. Their mother, Ruby, who was quite young and pretty, died far too soon.

Frank was probably my favorite friend back in Bear Swamp days. He had a keen mind, and was always friendly and dependable. He remained in the community and married a young woman, Velma Bryson.

Frank worked in the textile industry until his retirement, and his wife Velma bought and ran several ladies' clothing stores, which were quite successful. They still live in the community where Frank grew up. They are the parents of two outstanding girls.

Virginia, Frank's sister, was also a very fine girl, and has made a wonderful mother for her children. If she had been a few years older, I would have asked her to be my girlfriend. Virginia and her husband kept Morris in their home for years before he passed away.

Sophie Foster was a first cousin of my mother's, and we were in the same class at Blue Ridge School. She married Julius Addis, one of Earl's cousins. They produced a fine family and were devout Christians. During World War II, when I was in the army hospital in Thomasville, Georgia, Sophie's husband, Julius, was there too, and I rode home with him several times. Sophie and Julius spent all their lives in and around Walhalla.

Other schoolmates of mine at Blue Ridge School in the 1930s were Rubye Allen (now known as Robin), and her two brothers, Arthur and Olen. At that time, they lived in the old Phinney place, over on the Seneca-Walhalla Highway. Their parents, like mine, were tenant farmers, and although she did not have fine clothes, or anyone to show her how to make the best of herself, she was, nevertheless the prettiest girl at school, with natural blonde hair, blue eyes, and fair complexion.

And I remember how friendly and kind she was to me when we started to Walhalla High School. There was a vast difference in the social structure of the school. The country and mountain students who were bused into town were not accepted on the same social level as those who lived in town. I was not a very outgoing person at the time, and consequently, I was often alone.

I can recall walking from the school where the bus unloaded over to Kenneth Deaton's 5&10 cent store on Main Street, to buy candy. Rubye would walk with me both ways and hold my hand. That left a lasting impression on me. I really appreciated any friendliness shown to me back then.

In 1938, during my last year of high school, the Allens moved down the highway, closer to Seneca, thus putting them in the Seneca school district. Rubye attended both her junior and senior years at Seneca, where she became the most popular girl at school, and was chosen as Beauty Queen. What a vast difference in attitude between the two schools.

Shortly after finishing high school at Seneca, Rubye married a nice young man named Dudley, who had just graduated from Clemson. At his graduation in 1942, Dudley received a commission in the U.S. Army (infantry) and was immediately called into service. Rubye and Dudley had one child, a daughter, before he was sent overseas. He was killed in combat almost at the same time and in the same vicinity where I was severely wounded in the minefield in the Alsace-Lorraine area.

Rubye came back to live with her folks for a time, then took a job in Washington, and later married a Marine officer. They had several children and lived for quite awhile in Hawaii. Then, her Marine husband died while still quite young.

At the beginning and during the war, Rubye had worked for civil service and had achieved quite a good rank. After the death of her husband, she went to Saigon during the Vietnamese War and worked as a civilian, where she met Keith Adamson, who was with the diplomatic corps. They were married and now live in Honolulu, where they enjoy a very comfortable and active lifestyle.

I have kept up with Rubye through the years, both directly and through her brother, Olen. Ann and I have visited her in Hawaii on a number of occasions.

Olen Allen, Rubye's brother, went to Walhalla High School for a short time. He must have been about fourteen, when he left home. He ended up in Peoria, Illinois, and a nice family accepted him into their home, where he lived for several years while he worked for the LeTourneau Company, the same company where I worked, beginning in 1941. Then, Olen joined the navy, was sent to various schools, and he climbed to a high non-commissioned rank.

After the war, he returned home to West Union, where his folks were living at the time. Rubye was also living at home. Knowing that I had enrolled at Clemson, she came to me and asked if I could help Olen find some way to be admitted to Clemson, also.

I introduced him to my former high school math teacher, Miss. Karo Probst, who tutored him for several weeks. Olen took and passed the GED examination, which gave him the equivalent of a high school diploma. Then, he took the entrance examination to Clemson and passed with a high enough grade to be admitted into the School of Engineering.

For the next several years, while we were students, Olen and I became very close friends. Since he had a good car and I had none, we went many places together, sometimes with Rubye, and other times on dates. He was my best man when Ann and I were married in Senoia, Georgia on December 29, 1946. He was an extremely well mannered person, almost to the extreme, and also a perfectionist.

Olen finished Clemson with a high class rank and went to work for the DuPont Company. Later, he was with an electrical contracting firm in Myrtle Beach, South Carolina.

Olen married a distant cousin of mine, Sarah Crenshaw. They had several children and she made a very fine wife and mother. But after several years of marriage, he became manic-depressive. That plagued him for a number of years, and when he retired, he moved back to his parents' home in West Union.

While visiting a distant cousin up in the mountains of Virginia, he had a fatal automobile accident on the treacherous mountain highway. I did not learn of this until a few weeks after the funeral.

Rubye, Olen, and I were good friends for years, beginning when we went to school together. I give thanks every day that the three of us escaped the harsh life of a sharecropper.

Chapter 13

The Backcountry Elite

To understand the section of the country where I grew up, you have to go back to some of the early history of South Carolina.

The state is made up of three distinct sections, the low country, the upcountry, and the backcountry—that small mountainous area in the northwest corner, that is part of the Blue Ridge.

Not only the topography divides the state. Each section has its own culture, its own history. The small yeoman farmers, mainly Scots-Irish, of both the upcountry and backcountry, did not own a number of slaves as did the great coastal plantations, although some of the wealthier landowners had a few, prior to the Civil War.

Many of my mother's ancestors are buried in this church cemetery, since both the Whites and Sharps were involved in the early establishing and daily operation of the church.

Most prominent was Alexander White, who fought as a young man in the Revolutionary War. Because of his service, he was given a land grant in the Tamassee area. My grandfather, Wesley White, apparently attended this church when he was a child. When the family moved west of Walhalla to Negro Fork Creek, he joined Zion Methodist Church. Located just off the Walhalla-Highlands Road to the north, and across the road from Rocky Knoll Baptist Church, Zion Methodist is still in existence, but rarely used.

Besides churches and schools, new towns were also established. The names of these new towns usually reflected the heritage of the people who settled there.

The town of Walhalla was founded in 1850 by a colony of German immigrants, who had first settled in the Charleston, South Carolina area. The name was taken from Germanic mythology. Valhalla was a miraculous great hall of gold where heroes feasted, after doing battle.

One of the first acts of these settlers was to build

the town's St. John's Lutheran Church, which is still a magnificent structure.

My ancestors, both Stegalls and Whites, did not play a significant role in the town's history, although they were well-respected country people. Most of Walhalla's early history deals with these prominent German families, whose descendants are still known in the area.

As a child, I remember that one of the landmarks of the town was the elegant Biemann Hotel, built on the highest point of Main Street, in the old downtown area near the W.S. Moore Clothing Store. Set back about one hundred yards from the street, the hotel was three or four stories tall, German in architectural design, with its grounds beautifully landscaped. It was torn down in the 1930s, but I still remember it as a beautiful building.

The first automobile in Walhalla was a Maxwell, owned by C.R. Houckins, and used by the Biemann Hotel. This is where one of my mother's cousins comes into the picture.

The only person in Walhalla who could drive the automobile was L.O. "Ozzie" White, a first cousin to my mother. People say that Ozzie was quite a dude driving up and down the long main street of Walhalla in his horseless carriage. Ozzie later taught Bishop Cleveland, a black man who worked at the hotel, to drive, and Ozzie went on to better and more interesting things. Bishop Cleveland also had a connection with Bear Swamp. He was part of the black family that homesteaded and lived down in the Double Springs community in Bear Swamp.

The lore of any community is made up of both good and evil deeds, of men and women who rise to great spiritual heights, and the few who bring shame and dishonor to family and community. Walhalla was no different. Yet, we seem to remember the more violent times, rather than the peaceful ones.

An infamous day in Walhalla's history was in April 1930, when a lynching took place, and its outcome had a devastating effect on the peace and harmony of the surrounding area.

At that time, the population of Walhalla was about five percent black. One young black man, Allen Green, was a general handyman for a number of white families. At one of the homes where he was working, a woman accused him of rape. He was immediately arrested and locked up in the county jail in Walhalla.

A few nights later, a mob of white men showed up

at the jail and demanded that Allen Green be released to them. The sheriff, who lived on the first floor of the jail building, resisted, but he was knocked unconscious with a blow to the head. The mob leader took the key, went inside the jail, and pulled Green out.

The mob carried the black man from the jail, through town, and on through West Union to a spot near Burns Mill on Cane Creek, where he was tied to a tree. One man in the group, belatedly in sympathy, finally shot him in the head to end his suffering. Then, the entire mob riddled the body with shotgun, rifle, and pistol fire.

Several days later, Green was buried at the Flat Rock Baptist Church, a black congregation on the southwest side of Walhalla. To avoid further violence, the hearse and procession passed quietly by our house at the DuBose place, rather than going the more direct route through West Union and Walhalla.

I was nine years old at the time. I feel sure that I knew, or certainly my dad knew, many of those involved, but I never heard a name mentioned. No one was ever prosecuted. No investigation, no warrants, no indictments were ever made. Later, there was some question as to the integrity of the woman who had made the accusation.

The 1930s in Oconee County were difficult years in our history, when many people did not have enough to eat, when jobs were scarce, and when communication was inadequate.

But it was also a time when a number of men and women showed acts of kindness and generosity. They helped the poor, encouraged the timid, and reflected the beauty of forgiveness. In my estimation, these unnamed citizens also became the backcountry elite.

Chapter 14

High School Days in Walhalla

I have been back to only one high school class reunion, the 50th. Times had not changed from the years I was a student at Walhalla High School. There was still little communication.

As I look back to those earlier days, when I was a little country boy, with few friends in my class, I realize that it was mostly my fault. I was shy, and when the city folks seemed not to be interested in a boy from Bear Swamp, I kept to the friends that I had already made, such as Frank Lee, Lloyd Addis, and Rubye Allen, who were not even in my class.

But I remember a few teachers who helped me tremendously. The first was our agriculture teacher, Doc

Poore. He had served as a veterinarian in World War I. He was a fine old gentleman, although the boys gave him fits. He was one of the first teachers who indicated that I was not dumb.

Another teacher who helped my self-esteem was my math teacher, Miss Karo Probst. She had the reputation of being very strict, and it was hard for any student to pass in her class. The first "A" that I ever made in school was in her Algebra class. Fifty years later, I finally located her in a retirement home in Phoenix, Arizona, to tell her how much I appreciated having had such a fine teacher.

One of the highlights of my life back then was as a member of Future Farmers of America and going away to summer camp. I went a total of six times, four in high school and two when I was a student at Rabun Gap, returning home to go with the old gang.

My favorite place to camp was Bluffton, in the southeast corner of South Carolina. There we lived a very primitive existence, taking our own live chickens and other foodstuffs that would not spoil, since we had no refrigeration. My first glimpse of the ocean was at Tybee Island, near Savannah, Georgia.

At this time in my life, I wanted to learn all I could about farming, so that I could be the best. With my dad's cooperation, I had some prize-winning FFA projects in cotton and forestry.

I don't know why, but I had an incessant desire to go on to college after high school graduation in 1938. I am sure that this idea came from my mother and father and other loved ones.

The only college I knew about was Clemson College, a short distance away. My agriculture teacher and I worked hard, trying to find some way of coming up with the money—four hundred dollars, including books and uniforms. I could get a scholarship for $100, and work in the dining room for another hundred. But that left two hundred dollars, at a time when the family annual income was only $250. I got catalogs from all of the colleges in South Carolina, but when we got down to the money, the dream of a college education ended.

Then, in the fall of that year, a young lady named Helen Barrett came to teach at Blue Ridge School. She told my mother about Rabun Gap-Nacoochee Junior College, a farm school where a student could work for his or her tuition. By then, it was November, and the school year had already begun. But I sent in an application, and I was accepted.

My dream had suddenly been reborn.

Stegall (far right) with Junior College friends

Chapter 15

College Days

The fellowship that I found at Rabun Gap-Nacoochee Junior College was exactly opposite from my experience in high school. Some of the dearest friends that I have had all these years were people I first knew at Rabun Gap.

During the ten years that the junior college was part of Rabun Gap-Nacoochee School, about six hundred men and women attended. We were all very close, almost like family members. This is especially true of those in the same class, those a year behind us, and those a year ahead of us. We worked, studied, and had fun together, establishing meaningful friendships and memories that have lasted a lifetime.

Yet, those first few weeks in a totally different environment were difficult ones for someone who had left home for the first time, not including those brief trips with Future Farmers of America.

It was November 15, and the fall term had already been in session for almost a quarter, when Helen Barrett, the Blue Ridge teacher, and Lawrence Barker, who later became her husband, drove my mother and me to Rabun Gap, twenty-five miles away.

With no money at all for tuition, I was told that I could work for the rest of that quarter, to build up tuition credits, so that I could go to class the following quarter.

I had never seen the school before, and I didn't know a soul there. That first night, as well as a few others, I was so homesick that I cried myself to sleep. Two weeks later, Thanksgiving holidays came along. Although I had been working such a short time, I was extremely anxious to go home and visit with my folks.

I now realize the courage and guts it took to return to school to work outdoors during the winter quarter. On many days the temperature was about ten degrees. I plowed, cut timber, snaked logs back from the woods to the campus, and chopped firewood for the big furnaces.

When the authorities discovered that I could handle a team of mules, that became my job, doing whatever a team of mules was required to do. At the end of the winter quarter, I had earned enough credits to enroll for class. This was really a treat for a number of reasons.

We went to school four days a week, from 8:00 A.M. to 5:00 P.M. and on the two days we worked, the hours were the same. We had twelve hundred acres under cultivation, so most of the work was farm work, which I had learned at my dad's side.

By early spring, the baseball season started. So in addition to working and going to classes, we also had team sports. Baseball was my first love. The coach was also the basketball coach, and although I did not play basketball, the coach had me dress out with the team to scare the opponents. I was six feet, six inches tall.

Because of my height, I played first base on the school baseball team. I was a good hitter, but my worst feature was that I was slow. We played other junior college teams throughout the state and, in the summertime, we teamed up with a county team in Clayton to play in a textile league. When we had a baseball game, we got credit for working, even though we were playing sports.

After having gone to junior college for seven quarters, I graduated on May 29, 1941. I remained that last spring to play baseball. When I graduated, I still owed forty dollars for tuition. For the first time, my dad was able to help me financially. That forty dollars was his total investment in my college education.

Besides earning tuition working on the school farm, I also took on extra jobs, such as firing furnaces, assisting in the woodworking shops, and privately, serving as a school barber.

For the first year I had competition in cutting hair. The older student charged fifteen cents per head, so I felt that I should charge only ten cents. But after he graduated, I went up to fifteen cents. If I cut four heads of hair a night for five nights, I received sixty cents a night or three dollars for the week. That was far more money than almost anybody had at school. I now marvel at how my roommates put up with me in that tight little room where they were trying to study.

I roomed with Foster Goolsby for a short time when I first went to Rabun Gap. He graduated a year before I did, taught school for a year, and then worked for R.G. LeTourneau before going into the U.S. Army Air

Corps. He was a much-decorated pilot, receiving the Distinguished Flying Cross for his bravery in World War II. He married his Rabun Gap girlfriend, Frances Taylor.

Through the years, we have remained very good friends, seeing each other at school homecomings every year. Many, many times, we have gotten together on golf outings in the mountains, at Gulf Shores, and other places. He and Frances live in Fitzgerald, Georgia, where he has tried several times to retire from the educational field. But each time, he has been persuaded to return. At present, he is serving as Special Consultant for the Georgia Accrediting Commission.

Elbert Bishop of Campobello, S.C, was one of my roommates for two years. We graduated together and both intended to go on to school at Berry College. However, the war came along and we both ended up in the army. After the war, we both went to Clemson College, where Elbert received both a bachelor's and a master's degree in agriculture, and I got a degree in engineering. Because of our dramatically different courses, we did not see each other often. Elbert married a fine girl, Sarah, and they have several outstanding children. He is now retired as a county extension agent and enjoys golf.

Edwin Bishop was a year older than his brother, Elbert, although they were in the same class at Rabun Gap. After graduating, Edwin went on to Berea College in Kentucky, where he received his first degree. When the war came along, he was an outstanding flight officer.

After the war, he received a master's degree from the University of Kentucky and a doctorate from the University of Chicago. Edwin has held academic positions as Vice President of the University of North Carolina at Chapel Hill, President of the University of Maryland, Chancellor of the University of Arkansas, and is President-Emeritus of the University of Houston. He married Dorothy Anderkin of Kentucky, and they have three outstanding children. Ed and Dorothy live in Chapel Hill, North Carolina.

My favorite girlfriend while at Rabun Gap was Linda Wilson. Actually, we were friends more than sweethearts. She married another roommate of mine, Morris Ramey, who also was with me in high school. Morris graduated from Furman University and Southern Baptist Seminary. While serving as minister at a church in Gainesville, Florida, Morris developed cancer and lived only a few weeks after diagnosis. Linda and Morris had two fine children, a boy and a girl. Years later, Linda married an older man, Allen Hasty, a fine Christian gentleman who was involved in missionary work. He lived to be one hundred years old and died a few years ago. I usually see Linda at Rabun Gap homecomings and also keep up with her through her sister, Louise, who has also become a very good friend since leaving Rabun Gap.

Math teacher, athletic director, dormitory manager, and best of all, my baseball coach, Mr. Jack Acree understood our problems and helped us in many different ways. He came from the same financially strapped background as the rest of us.

We both left Rabun Gap about the same time, May 1941. He became the first director of the LeTourneau Machinist School in Toccoa, Georgia. I joined him there as a student a few months later.

We have kept in contact through all these years. He married Ruby Taylor, another teacher at Rabun Gap. After having two fine daughters, Ruby developed cancer and passed away. After several years, Jack married his present wife, Jackie, and they make a very fine pair. Jack has been an educator all his life and is in the Georgia Educational Hall of Fame. He loves golf, but did not begin to play the game until after he was sixty-five. He is approximately eighty-seven years old.

Other Rabun Gap alumni that I remember vividly are: (1) Gibb Gosnell, (2) Jake Griffin (deceased), who married Ruth Robinson, (3) Julia Head, (4) Talmadge McNeal (deceased), (5) Pauline Moore, (6) Wesley Williams, (7) Harding Cain, (8) Rip and Rogers Chandler (both deceased), twin pitcher and catcher on our baseball team, (9) Sue Coleman, who married Rogers Chandler, (10) Joe Martin, retired professor from University of Tennessee, who married Mary Addington; a special friend from Clemson days, (11) Helen Barrett, a 1938 graduate of Rabun Gap, who told me about the school.

Clemson College Friends

Other college friends include two from my days as a Clemson student after the war. They are Frank Hawthorne and Alex Moss.

Frank was my first roommate in the Veterans Barn. He studied textile engineering, and after a successful career with Greenwood Mills, is now retired. He and his wife Verna have three children, a daughter who is a nurse, and two sons, who hold the honor of being the number one and number two all-time engineering students at Clemson.

Frank and both sons are low handicap golfers.

Alex studied civil engineering. He taught at Clemson for several years, then went into the pre-stressed concrete business, before moving into general contracting. He and his wife Ellen live near Williamston, S.C, and are members of the Pisgah Baptist Church, with my cousin, Ruby Couch.

From my first taste of higher education at Rabun Gap-Nacoochee Junior College to my engineering degree at Clemson College, I shall always be grateful for people like Dr. and Mrs. Andrew Ritchie at Rabun Gap and those in charge of the GI Bill, who had the vision to help students like me achieve success. And I am eternally thankful for the friends that I made along the way.

Stegall in army uniform, with his mother Weedie.

Stegall at Clemson College in 1946
Veterans wore their own military uniforms

Chapter 16

STEGALL AND WORLD WAR II

World War II started full blast in Europe in 1939. America began an unofficial buildup at this time, and by 1940 was drafting people for military service.

In the spring of 1941, I finished Rabun Gap-Nacoochee Junior College and then worked and played baseball for several months at LaFrance Industries. Then, on July 29, 1941, I entered the LeTourneau Machinist School in Toccoa, Georgia. A year later, I was transferred, along with thirty others, to the new Vicksburg, Mississippi plant that was under construction. I graduated from the machinist school in July 1943, and then worked as a full-fledged machinist until May 1944. Normally, I would have been drafted in 1942, but was deferred because the LeTourneau Company was engaged in making war materials, such as shell casings and heavy grading equipment.

By that time, many of my friends had joined the service of their choice. Their particular interest was in joining the U.S. Army Air Corps, later called the U.S. Air Force. I also tried to join both the air corps and the navy. There was a rule at that time that a man could not be taller than 6'4" in either of those two services. I passed the physical for both, except for the height requirement. I was 6'6"—two inches over the maximum height. Disappointed, I even put in an application to join the Merchant Marines, but had second thoughts and did not follow through.

In May 1944, the U.S. government declared that there would be no more deferments for LeTourneau, and all the single guys, as well as several married men, were called into service. Some of us from the Vicksburg plant had our pre-induction physical and orientation at Camp Shelby in Mississippi. We stayed only two or three days, and still had our choice of the army or the navy. We all chose the navy, but Howard McNeal and I were turned down. Once again, I was considered too tall. We both went into the army, since there was nowhere else for us to go.

I left Vicksburg the latter part of May to visit my folks in Walhalla and also to be there to report to the draft board and then on to the army. The last night I was in Vicksburg, Mom LeTourneau volunteered to loan me her car so that I could spend my final evening with my girlfriend at that time, Margie Marks.

After a few days at home in West Union, I reported to the draft board in Walhalla, where I was loaded onto a bus bound for Fort Bragg, North Carolina. The actual induction day was June 1, 1944, five days before the Normandy invasion. For one week, we took all sorts of tests, evaluations, and shots. It was a "hurry up and wait" kind of week.

While lining up for supper at the mess hall one evening, we were told to step aside and let a group of combat veterans, dressed in heavy winter ODS, ahead of us. As I watched these veterans line up, I recognized a dear friend of mine from Rabun Gap days—Joe Craft, a fine Christian. He was a couple of years ahead of me in school, and someone that I greatly admired.

Oddly enough, when I asked my sergeant if I might have some time off the next day to visit with my old buddy, he gave an affirmative answer.

So Joe and I spent the afternoon together. Besides discussing old times, I was interested in what he had to say about his 34th Infantry Division combat fighting through North Africa, Sicily, and Italy. That afternoon was the last time I saw Joe. Based on his overseas duty, age, and combat experience, he was among the first of the GIs to be rotated home under the point system. He later had a successful career in education at Hartwell, Georgia.

Joe had married a very fine girl, Ouida Roberts, who had also attended the junior college at Rabun Gap. But I never saw him at any of the school reunions, and I learned recently that he has died.

When our induction procedure was completed at Fort Bragg, we got on a Pullman train and headed to Camp Blanding, Florida. With no air conditioning, we had to leave the train windows open. By the time we arrived at Camp Blanding on June 7, our bodies and clothes were covered in soot and smoke from the old steam engine.

Through June to September, we spent seventeen weeks of combat training, which was the most rugged endeavor I have ever gone through. On most of the days in the so-called shade of the skimpy pine trees in that central section of Florida, the temperature was 112°. The training company started out with two hundred men. Many of the draftees were older. Those from the north simply could not take the heat. After seventeen weeks, the company had dwindled to one hundred men.

On our first day at camp, we did routine exercises and pushups. I was not able to do even the first pushup. Embarrassed, I was determined not to give up. On the last day at Camp Blanding, we had Olympic type contests. I was pleased to learn that I had done more pushups than anyone else in the company. Also on that last day, we had a five-mile run with rifle and full pack. Earlier, because I was so slow, I had been given the nickname of "Lightning." I won that five-mile run, too, since it was more a test of endurance, rather than speed. I am sure that the training and conditioning I received helped me survive the wounds received in Europe the following March.

While at Blanding only a few weeks, we had been allowed to make application for officers' training. My IQ grades were good, so I applied. I made it to the first interview, but was not selected. At that time, they were accepting virtually no applicants, since the need was for infantry replacements. Only one man out of the entire battalion was selected. He was a graduate of Clemson College with four years of ROTC training, but for some reason had not received his commission at graduation.

My training at Camp Blanding was with an anti-tank company. The only difference between it and a rifle company was in learning to use 37mm anti-tank guns. Mounted on wheels similar to a howitzer, they were about as effective on tanks as a popgun, and were considered a very sad joke. When we got overseas, we were given training with 57mm anti-tank guns. They were a very small improvement, since the Germans were using 88mm cannons against us.

About the middle of October, we were given travel fare and orders to report to Ft. Meade, Maryland, after a ten-day terminal leave, prior to going overseas.

Immediately, I hitchhiked to Vicksburg to visit my old buddies. When I got there, I was disillusioned. Virtually all of them had gone into service and my best girlfriend had married her former fiancé, Mac, who was a major in the air corps. Later, he was reported killed in action, but he turned up as the Allies captured Rome.

From Vicksburg, I went to Walhalla, or rather West Union, and visited my parents and other loved ones.

Then, I caught the train for Baltimore, Maryland at Seneca, S.C.

At Ft. Meade, there were a number of German prisoners of war. We also had quite a few prisoners of war at Camp Blanding. But the difference between the two groups was astounding. The ones at Blanding were well disciplined and seemed to be pleased to be in Florida and away from the war and chilly temperatures in Europe. The ones at Ft. Meade were the most belligerent and arrogant group I saw anywhere during the war.

For several days, we stayed at Ft. Meade for more orientation. Since our evenings were free, I went into Washington to visit with another old girl friend, Phyllis McFarland, whose mother was a friend of Mom LeTourneau. This was a good experience, except that I smoked back in those days, and Phyllis and her mother strongly objected. My cousin Harold Stegall had gone into the service several months before me. He was stationed in the naval barracks near Washington, so I spent the night with him.

After the red tape was completed at Ft. Meade, we traveled by train a short distance to Camp Patrick Henry, near Norfolk. Knowing that we were on our way to a combat zone, my buddy at the time and I went to worship service each evening, where we heard excellent sermons and received communion. Then, three thousand of us shipped out to Europe.

We were in a convoy with other ships and were escorted by two or three destroyers. In an attempt to avoid German submarines, we traveled in a zigzag course.

On our ship was a group of air corps officers, who said they were on the way to the China, Burma, India Theatre. So we thought we were going there also, instead of Europe. We sailed through the Strait of Gibraltar and on through the Mediterranean Sea toward Italy.

After fourteen days of sailing, we woke up one morning with the ship anchored in the center of a beautiful bay. We could see the mountains almost surrounding us, with the ridges of pine trees resembling umbrellas. We first mistook them for palm trees, but when we looked farther east, we could see smoke or steam rising from Mt. Vesuvius. Then, we realized that we were in Naples, Italy.

Since the harbor was badly torn up and the ships could not reach the docks, we were transported to shore in small boats. Debris was in every direction. Kids were living in boxes and trash heaps. They were begging for chocolates, cigarettes, and anything else we had. Perhaps most shocking of all, some of the kids were soliciting for their sisters.

I believe we arrived in Italy on November 12, 1944. This was the first time that I had ever set foot on foreign soil. Trucks picked us up and we rode some sixty or seventy miles north through the town of Caserta and on into the Volturno Valley, where a large replacement depot was located.

For two months, during Thanksgiving, Christmas, and New Year's, we stayed at the depot. We had a special meal on all three occasions. We also had special training with 57mm anti-tanks guns, bazookas, the new machine gun, and other weapons. Sometimes, they took us out on long walks, sat us down, and lectured us on various subjects, some having to do with war, others just to kill time.

In a muddy field that had been cultivated the previous summer, we lived in ten-man tents and slept on fold-up canvas cots with wooden legs.

Although this was southern Italy, the temperature seemed to hover around 40°, or just above freezing. We could see the snow on the high mountains and feel the cold wind that blew down from the mountains. The scenery was beautiful, but we had no heat, and inside our tents, we were ankle-deep in mud. To stay warm in our cots, we had to make a sleeping bag of our blankets.

In the evenings, we would walk about a quarter mile or more through the mud to an old dairy barn that had been converted to a USO facility. There, we would have coffee, doughnuts, and conversation.

Our canteens were made of aluminum, with a thick lip around the top edge. When we filled the canteen with hot coffee, the liquid would get cold before the aluminum rim cooled enough for us to bring it to our mouths.

The USO was the only recreation that was available, so we made the most of it. Occasionally, some in the group would try to furnish entertainment. But we all drank so much coffee, that a few hours after going to bed, we had to go to the latrine.

There were guard stations all around the tents, to make sure that we made it all the way to the latrine, which was actually a slit trench. Most of us were in such a hurry, that we would not lace our boots all the way. The strings would get coated with mud and were an unholy mess to deal with the next morning when we dressed.

On one occasion, we were scheduled to make a

twenty-five mile hike north to Cassino, site of the monastery that had seen such stiff German resistance, causing many casualties among the Americans and Allies. It had finally been blown to bits several months earlier by the air corps. On the day of the long hike, my name appeared on the list for KP duty. Instead of walking the twenty-five cold miles, I worked indoors where it was warm, and I had plenty to eat the entire day. That was the only time I enjoyed being on KP duty.

The 10th Mountain Infantry Division had been fighting the Germans in the Po Valley in northern Italy. This was the division we had been expected to replace. But before we were called out, the Battle of the Bulge occurred around Christmastime. About the middle of January, some three thousand of us were transported back to Naples, where we boarded a British ship and sailed for Marseille, France.

It was not a luxury cruise. I was halfway sick the entire time, although not ill enough to lose the one meal of the day of stew or thick soup, which we were served from a giant kettle.

Early one cold evening, we arrived at the docks in Marseille, where the drizzling rain was mixed with snow.

As was often true in the army, we stood on the docks and waited for someone to come along and show us the way.

While we waited, a crane was busy unloading the nets filled with merchandise from the ship. One of the nets broke, dumping all contents onto the dock. At first, we figured that the goods were ammunition, rifles, medical supplies, or food. Instead, it was a load of Beef Eaters English whiskey. When the GIs saw what it was, they all made a dive to get a bottle of this scarce commodity. Many were successful, and for days, there were plenty of drinks for anyone wanting whiskey.

Finally, a lieutenant came along and had us line up to follow him. It was pitch black and drizzling rain. We walked for what seemed like hours through the streets of Marseille. By that time, the lieutenant had enough sense to realize that we were going in the wrong direction. We about-faced and retraced our steps. We found the train not far from the place where we had docked.

I was in a compartment with several other men all with full pack and rifle. We had no room to move and we no way to make ourselves comfortable. The farther north we went, the colder it became, with snow and knocked-out German guns and tanks visible from the window.

We soon discovered that we had the lowest priority of anyone traveling by rail that night. Our train would sidetrack, stop, and wait for every other train on the track. Coming ahead of us was a train carrying the wounded, or ammunition, or food. After we had sidetracked several times, some of the GIs got up enough nerve to jump out the windows and doors and head for the nearest farmhouse. When the train was ready to go again, the GI engineer would toot the whistle and everyone would come running. Many brought back a big loaf of French bread or a bottle of wine. The wine was apparently without sugar and the alcoholic content was low, but it was good enough to keep us halfway warm. So on the way from Marseille, we ate whatever we could get our hands on—C rations, "10 in 5," and anything we could beg from the French.

I recall one day when I badly needed a shave. I went up to the steam locomotive and got permission from the engineer to fill my helmet with hot water. It was more than enough for a fresh, clean shave. I felt much better after that.

We finally arrived in Strasbourg, France, close to the German border, after five nights and four days. If we drove that distance today, it would take us four or five hours. We unloaded the train and marched into a large four-story building that had been a factory. Realizing they were going to be overrun by the Allies, the Germans had taken all the machinery into Germany, and the building was bare. Wrapped in our blankets, we slept on the floor.

I had developed a severe throat condition, and I could hardly speak above a whisper. When the sergeant came in the next morning, he said there was a field hospital in Strasbourg, and asked if anyone needed to go on sick call. Although I did not feel as bad as I sounded, I thought I had better take advantage of getting some medication. So I had the sergeant put my name on the list.

Shortly after that, another sergeant came in. Out of 3,000 men in the group, he called the names of six people, who were to fall out, with full pack and rifle. I knew three of them, who were machinists. In the army, it did not take long to become attached to someone who had a common interest. I figured that we were not being called out to take the place of some poor infantry combat soldiers who had just been killed. I decided that they needed machinists to rebuild tanks or other machinery, possibly in some place like Paris.

I had the sick call sergeant scratch out my name,

and I fell out with the other men. We were told to get into the back of a truck that was covered with canvas. That canvas was the only protection from snow blowing in all directions. We drove for a couple of hours, about fifty or sixty miles, and all that time, snow was blowing into the back end of the truck. By the time we stopped, our uniforms and all our equipment were covered in snow.

When we jumped out of the truck, we landed in the most beautiful little Santa Claus-looking village we had ever seen. Of Alpine architecture, it was covered in snow-the ground, the vegetation, the houses. It was the most peaceful view you could imagine.

A colonel and several other officers came out of a former dwelling house. They shook our hands and invited us inside to have coffee or hot chocolate. I thought for sure that we had gone to the wrong army, because no officer had ever treated us this way. After warming up before a nice fire and enjoying the hot chocolate, we were told that this was the 121st Cavalry Reconnaissance Squadron.

The machinist connection had been associated with the fact that this was a mechanized unit, and they wanted people with some mechanical aptitude. It did not make sense exactly, but hardly anything made sense in the army.

We were in Ludweiler, a few miles into Germany and not too far from Saarbrucken. Half the town was controlled by the Germans, and the other half controlled by us. We were not going to Paris, after all. Instead, we were on the front line.

Cousins Bill Stegall and Spec. Jamison with W.C. Stegall

Chapter 17

The Front Line and Beyond

From being trained in the heat of central Florida for combat duty in the South Pacific, we were now knee-deep in snow in Germany.

As part of Troop "B," Third Platoon, I stayed in Ludweiler until after the middle of February. Lt. Martin from Texas was in charge of the platoon. The staff sergeant was named Vaith, and the sergeant of my section was Tom Klump. He and his twin brother James were both sergeants in charge of two different sections or squads. Little fighting was going on at the time, although the platoon sent out patrols and got in a firefight once or twice. I was not in one of these patrols.

Our section, or squad, had the responsibility to

man an armored car, which had been dug in and camouflaged on top of a hill, with surrounding foxholes, where we could watch the Germans. Normally, there were four men on duty at all times. Occasionally, we relieved the watch in the downtown area. From there, we could see the Germans going and coming from their field hospital, and hear the nightly train bringing in supplies to them.

Occasionally, the enemy fired mortars at us, more for harassment than for any harm they could do. The Germans apparently thought that we were occupying a heavily built school. They almost destroyed the school with their fire, while we were in our foxholes or in houses under the slope of the hill, where they could do little damage.

One day we found some sleds, and while off duty in the combat zone, we had a bantam (jeep) pull us to the top of a hill, and then we rode the sled downhill into the town. This was great fun, but may have been dangerous, since we didn't know whether some Germans might have been hiding in one of the buildings we passed.

On several occasions, we went exploring around downtown, to see what we could find in the empty buildings. Suddenly, we heard a loud noise that sounded like German heels stomping in the next room. We froze.

Before we could get to our hand grenades to throw, a big billy goat stuck his head around the corner.

Ludweiler was the most comfortable place we had found since leaving home. The Germans had left much of their foodstuffs when they departed, so we had plenty to eat. In many ways, that was an enjoyable time. When we left Ludweiler for Merlebach on February 11, our group was relieved by the 101st Cavalry Reconnaissance Squadron. After the war, I discovered that a dear friend of mine from Bear Swamp days, "Punk" Lee (who married Rachel Addis) was a sergeant in that squadron. He was killed a few weeks later.

In Merlebach, which was just over the line in France, we stayed for about a week to rest and to train on new equipment. For more than a hundred years, that area of France had been fought over, belonging at times to France and then to Germany. Most, if not all, of the natives spoke German.

Here, we had excellent sleeping facilities in people's homes and a nice cafeteria that we turned into a mess hall. A large coal mine had installed banks of hot showers, which we enjoyed. We also found a French lady to wash our clothes. Her husband was German and in the army.

When I first joined my outfit, I was supposed to have been a gunner on board an armored car. But the military soon realized that my 6'6" frame would not fit into this space. So I was assigned as a bantam jeep driver for Sgt. Tom Klump. In this new position, I was promised a corporal's stripes, but they never caught up with me.

From Merlebach, we drove several miles south to the little rural village of Eyweiler. The barns were joined to the houses, and in turn, all houses were connected to each other, like row houses. In front of each house stood a large pile of manure, taken from the barn. The bigger the manure pile, the more affluent the people were thought to be.

Again, we stayed in homes. Three of us had beds in an attic to sleep in. One of the men was Eldred DeKoster, who later became a sergeant. The owners had stored apples in the attic, which we made good use of. Then, the school in the middle of the village was turned into a mess hall, where we ate good, hot chow.

We spent two nights in that attic room. By noon of the next day, we enjoyed a hot lunch at the schoolhouse. That afternoon, we were assigned to visit some of the old Maginot Line pillboxes. It was not clear whether this assignment had been given, merely for something for us to do, or whether a few Germans might actually be hiding in some of the pillboxes.

The French had built the Maginot Line in World War I, with guns pointing toward Germany. In turn, the Germans had built the Siegfried Line, with guns pointing toward France.

As we were leaving the village, I was invited to ride with several other GIs in a half-track. That is the last thing I remember of that day, March 5, 1945. Later, I was told that we drove some two miles or so, parked our vehicles, and walked across the field to get to one of the pillboxes, where we would have our lesson. We were always having lessons and lectures on some new method of fighting.

Snow still covered the ground, with a few visible patches where it had melted. The field looked as if it had been cultivated the year before, prior to the Germans mining it with 12 lb. anti-tank mines (12 lbs. of TNT).

Before the Germans left, the main road had become impassable because of a large bomb crater, so the field had been used as a detour around the crater. Knowing that we would have to take the same detour, they had mined this particular section of the field. The Germans made a habit of booby trapping one or more of their mines, and it was easy digging to hide them in the soft ground.

Before we arrived, the main road had been repaired, so our vehicles were not at risk.

Six of us, including Sergeant Jim Klump, began walking from the half-track to the site of our lesson. We had not gotten far when the man directly behind me tripped a hidden booby trap and two of the mines exploded. Three men were killed on the spot. The fourth, Sergeant Klump, suffered a skull fracture and concussion; the fifth, an even more grievous wound.

Three days later, I woke up in the 54th Field Hospital in Drulingen, France. Because of burns from the mine explosion, I was wrapped like a mummy, with only my mouth and nose sticking out. Many of my teeth were broken or chipped. Shrapnel had lodged in a number of places, including my elbow, which had disabled my right arm.

The hospital had been an elegant old manor house, with murals on the ceiling. When I opened my eyes, the first thing I saw was a beautiful young nurse dressed in white. She was standing on a chair, adjusting the white, fluffy curtains on a window. At first, I thought I was in heaven, with an angel hovering over me.

As soon as she realized that I had regained consciousness, she offered me coffee. It was served in a big cereal-sized bowl with plenty of cream and sugar, and I sipped it through a glass straw, so that I did not need anyone to hold it to my mouth. I never had such good coffee. I can still taste it to this day.

The nurse was Lt. Lavonia Hite from Evansville, Indiana. It was not long before we determined that we both had been to the same Christian camp, located at Winona Lake, Indiana. It was run by Mom LeTourneau and called Camp Bethany. She, of course, knew of the LeTourneaus and many others who were a part of this camp community.

On March 12, 1945, after a week at the field hospital, I was moved to the 23rd General Hospital in Vittel, France, thirty miles from Epinal.

The U.S. Army's 23rd General Hospital had taken over the LeGrande Hotel, a luxurious facility, which had also been used as a hospital in World War I.

The large ward, where I was placed, had been the

ballroom of the hotel. Even more impressive murals were painted on both the walls and the ceiling.

Adjacent to my bed was a full-blooded American Indian. On the other side was a black soldier. Across the aisle lay a captain from Due West, South Carolina, and in back of me, a German prisoner of war. The orderlies were Italian prisoners of war. They were a fun group and acted as if they might be on a picnic rather than at war. As you can see, the ward was quite integrated.

For the first three weeks or so, I was in traction and could not get out of bed. During that time, the doctors and nurses could not have been kinder.

I was amazed at the excellent medical treatment that had been established there in such a short time. But good medical treatment could not always work miracles. The German prisoner of war had an infected leg. No matter how hard they tried, the doctors could not save it, since gangrene had set in. The only way to save him was to remove the leg. As they came to take him from the ward to the operating room, he screamed and begged them not to amputate it.

When I was able to get out of bed and move around, I walked, with help, through the beautiful gardens surrounding the buildings. It was springtime. Flowers were popping up everywhere; swans were floating on the lakes; ducks were nesting, and birds were singing. I have many pleasant memories of that place.

Every afternoon, we listened to the Armed Forces Radio. Johnny Mercer was the disc jockey, and he played all of the popular songs of that time. His theme song was *Hampton Turnpike.*

After I was able to walk around freely, the nurses one day dressed me up as Little Abner and paraded me around the hospital. I was not aware that it was Sadie Hawkins Day. After being away from American girls for five months or more, it was a real treat to enjoy the beauty and friendliness of these nurses and other members of the medical staff.

In 1987, my wife Ann and I, along with Callie and brother, Charles, visited LeGrande Hotel. It was as beautiful as I remembered it, but the prices were quite high. When I mentioned that I had been there in World War II, the staff gave us free rein to go anywhere we wanted in the building and on the grounds. It was a unique experience for me to return, not as a patient, but as a tourist.

About April 17th, 1945, I went by ambulance to

the 43rd General Hospital in Les Milles, France. While there, I was evidently waiting for transportation to England, although I did not know it at the time. The army very seldom told us where we were going or why.

At this hospital, the orderlies were German prisoners of war. They were the exact opposite of the Italian orderlies at the previous hospital. I have never seen a more disciplined group. Their fatigues were pressed and clean as a whistle. When you spoke to them, they snapped to attention and showed a willingness to do whatever you asked of them.

After several days, many of us flew on a hospital plane to Diddington, England, to the 49th Station Hospital. As we were flying from France to England, the pilot announced over the intercom that President Roosevelt had died that morning at Warm Springs, Georgia. A little later, during the flight, the clouds broke, and as we looked out, we could see the English Channel and the white cliffs of Dover.

The hospital in England was housed in Quonset huts, with a great number of them located on land that had been a farm and pasture. The English people were quite accommodating. The young ladies were especially attentive, particularly on the weekends. They would bring cookies that looked pretty, but because of the wartime shortages, they had very little taste.

The men would come out and invite us to play golf. My right arm was still in a cast and there was some question as to whether I would ever regain the use of it. Of course, I was not able to play. I am not even sure that, back then, I even knew what *golf* was.

On that same trip in 1987, we also visited Diddington. I finally found the site of the hospital, but there was no sign that it had ever existed. But we had the privilege of meeting the lady who owned the property. She was quite nice and invited us back for tea.

About May 10, 1945, I left the hospital at Diddington. A number of us traveled by train all the way to Glasgow, Scotland. It was an all day journey through the beautiful countryside. Upon our arrival, we were told that we would soon fly back to the States. In those days, flying was considered something special. We were put into temporary barracks, and we expected to fly to the USA the next day.

The next day came, and everyone's name was called, except mine. Having become accustomed to the army routine, I did not protest or ask any questions. I assumed I would be on the next plane. I spent all that day there alone. Later in the afternoon, another group came in on their way back to the United States. By morning, the sergeant came in and read the names of those who would take the next plane. Again, everyone's name was read out, except mine.

By this time, I decided to take some action. I went up and talked to the officer in charge, and in a very impersonal army style, he wanted to know my name. He fumbled through his files for a few minutes, then came back and said, "You are already in New York."

I was able to convince him that was not true. He then seemed a little concerned, and I was on the next plane headed for the States. This was a twenty-four hour trip. We stopped in Iceland, where we refueled and ate. There was a softball game going on at twelve midnight, since it was the time of year when it was daylight almost around the clock. We also stopped in Newfoundland, and then went on to New York.

We arrived in New York on Mother's Day, May 12, 1945, and were given a free telephone call to our homes. My folks did not have a telephone in their little apartment in the basement of the Grover Crenshaw house. However, I knew that my dad often worked at a nearby store/filling station owned by Sam Albertson, and I had the phone number.

I placed a call to my mother at that number, and someone hurriedly drove down and picked up my mother and brought her to the phone. I had never talked to my mother by phone, and I did not recognize her voice. I was so concerned that I was speaking to the wrong person, that I placed a second call. I was finally assured that I was speaking to the right mother.

We flew to Atlanta the next day, to drop off several GI patients who had been assigned to Lawson General Hospital, where the Chamblee airport is now located. We then flew on to Thomasville, Georgia, where I was assigned to Finney General Hospital.

As I got off the plane at the Thomasville airport, the first thing I saw was the plane owned by R.G. LeTourneau. I immediately went over to the plane and looked inside. There was R.G's pilot, Red. I don't know who was more surprised.

"What are you doing here?" we both asked about the same time. Red told me that he had flown the old man down to make a talk that day. Then, of course, I told him my story.

Finney General Hospital was brand new, with excellent facilities, and well staffed by doctors and nurses. If the patient were physically able, it was customary for a GI returning from overseas to be given a furlough immediately. The next day an announcement came over the loud speaker that Private William C. Stegall was to report to Colonel Browne in his office.

I was both shocked and puzzled. What had I done? Why in the world would the Commanding Officer want to see Private Stegall? I washed my face and hands, combed my hair, and tried to look the best I possibly could. Upon arriving at his office, I was ushered in. The first thing to do, of course, was to salute the Commanding Officer. With my right arm in a cast, I was not sure what to do, so I saluted him with my left hand.

I still am not sure whether that was the correct protocol, but the colonel did not seem to mind. He announced that there were two ladies to see me. I looked around, and there were Mom LeTourneau and my mother. R.G.'s pilot, Red, had gone back to Toccoa and told Mom about seeing me. Mom then drove over to Walhalla, twenty-five miles away, picked up my mother, and the two had flown to Thomasville. When I met them in the colonel's office, arrangements had already been made for my furlough to begin that day. So I flew back to Toccoa with them.

My furlough lasted a month. During part of this time, I went back to Vicksburg, Mississippi. But I was disillusioned that even more of my friends there had scattered to the four winds, so I returned home to family and other friends.

When I went back to Finney, the doctors removed my cast and I began therapy on my right arm. Every morning I reported down to the therapy department, and a nurse exercised my arm in warm water. That nurse was "Stump" Cline's wife, whom I knew in Toccoa when I was with LeTourneau. It was good to run into someone with many mutual friends.

Earlier, at the 23rd General Hospital in France, the doctors had told me that they would give me the motion in my right arm that would be most useful, but it would probably be impossible to regain full use and movement. That was proved to be true. It became obvious at Finney, after a couple of months, that nothing much more could be done. I began hoping that I might be discharged from the army in time to enroll at Clemson that fall.

One of the great things that happened to me, while I was at Finney, was to receive a letter from my dear friend, Ann Welden. I got to know and appreciate Ann when we both worked at R.G. LeTourneau in Vicksburg. Ann had come out to visit her brother George, and while waiting to be admitted into nurses' training at Crawford Long Hospital in Atlanta, she had taken a job at LeTourneau.

In the plant, Ann became an inspector of 155mm shell casings, and our paths crossed many times during a day. She would listen to me as I told her all my troubles, especially with my girlfriend at the time—Carrie. Ann was a very pleasant, friendly, and attractive person. Except for a house party one weekend that Mom LeTourneau had arranged for a big group of us, we did not date.

When I received the letter from Ann, she was already in nurses' training at Crawford Long. As I traveled back and forth between Thomasville and Walhalla, I usually stopped in Atlanta. On one of these occasions, I went to see Ann. She was working and training at Henrietta Egleston Hospital for Children. She was still the same outgoing, friendly, and attractive person that

I had remembered.

While we were sitting on the porch in rocking chairs and enjoying the pleasant summer evening, I wanted to hold her hand. But I thought it best to ask her permission. She did not object. Later, as I got bolder, I looked in her eyes and said, "You have beautiful teeth. They're just like pearls."

"Pearl who?" she replied.

This should have killed any spark of romance, but it didn't. I learned to appreciate even more the pleasant girl that I had known in Vicksburg.

Back at Finney, I was in chow line one day. Next to me stood Howard McNeal, the friend who had also been turned down by the navy. We had both taken basic army training at Camp Blanding, where I saw him several times. He had gone on to Europe as an infantry replacement, joining his outfit in Belgium, just outside the town of St. Vith. While in a foxhole, trying to escape from a barrage of 88mm shells, he was severely wounded in the right knee. He spent almost two years in hospitals, and finally ended up with a stiff leg. Howard and I enjoyed each other's company while we were at Finney. His wife Evelyn would come to visit, and it was always a joy to see her, as well.

Julius Addis, a friend from Walhalla, who married my cousin, Sophie Foster, was also at Finney. He owned an old car with patched tires. We made a few trips back and forth in that car to Walhalla, but we always carried several spare tires with us. We had to change tires at least two or three times during the trip. The only tires that were available then were recaps.

Around the first of August, I asked my doctor, a lieutenant, if it would be possible to release me from the hospital and the service in time to attend Clemson College in September. He acted quickly on this and, in just a matter of days, I received my discharge papers. Dated August 7, 1945, it was the same day that one of the nuclear bombs was dropped on Japan.

My first night as a civilian, I spent in Atlanta. On the next day, I went on to my parents' little apartment in West Union. While in the army, I never received my corporal's stripes as promised. However, I did receive the European Theatre ribbon with two bronze stars, and a Purple Heart.

Shortly after I left Finney Hospital, Colonel Browne, Commander at Finney Hospital, retired and established a medical practice at Clemson, South Carolina, the home of Clemson College, where I would be a student for the next three years.

Ann and Stegall with first child, William
summer 1950

Chapter 18

Fifty Years of Business, 1948~1988

On February 1, 1948, I graduated from Clemson with a degree in mechanical engineering. I had been given college credit for some of my classes at Rabun Gap, as well as a few from LeTourneau Machinist School, such as metallurgy. So that helped in completing my degree a little earlier than expected.

While I was still a student, Ann and I were married. From that time on, I never made under an "A" in my classes. She was a good influence.

After graduation, we moved to Philadelphia, and on February 15, I started work at General Electric Company in their Switchgear Division.

After four months, I requested a transfer to the Air

Conditioning department, and we lived in Bloomfield, New Jersey until the middle of December. Cold weather had arrived by the middle of November, with the promise of more snow and a severe winter. Ann decided that it was time for us to move back to a warmer climate.

Then, too, my dad had died the previous June and my mother was left alone in South Carolina. So combined with the cold weather, the impersonal nature of working for such a large company, and feeling the need to be closer to my mother, we decided to move back south.

During the Thanksgiving Holiday, plus a few extra days, we made a quick trip to visit the GE Air Conditioning distributors in Raleigh, North Carolina, Atlanta, Georgia, and Birmingham, Alabama. All had shown an interest in hiring me.

After the war, new cars were almost impossible to buy, but paying a premium price of $1,800, we bought our first car, a 1948 cream-colored, two-seated Ford. It had no air conditioning.

That weekend, we drove in the Ford to all three appointments, with little time to sleep. All three of the firms offered me jobs, but oddly enough, the one that appealed to me most was the offer in Birmingham.

My only impression of the city had come from my traveling through earlier by bus, train, car, or hitchhiking. I had thought it was the very end of the world. I had not seen the fine residential communities, only the dirty, crowded areas with the steel mills and furnaces.

On this visit to Birmingham, I interviewed with Air Engineers, Inc., with a tremendously fine gentleman, George F. Holmes. He was a delightful person, warm and understanding, and his offer was equal to what I was making in New Jersey. After getting back to New Jersey, it took us only a day or so to decide that we wanted to move to Birmingham.

The only person we knew there was Blake Gambrell and his family. Blake was the son of Monroe Gambrell, whose place we farmed in South Carolina, in Bear Swamp. The Gambrells always spent their vacation back in their home place, and I had gotten to know their three children, C.B., Sarah, and Mary.

Mrs. Gambrell took us under her wing and showed us, on that first trip, how really beautiful Birmingham was, especially south of Red Mountain.

C.B., her son, was only a year or so younger than I, and he and his wife Olive, had been at Clemson part of the time we were there. On that trip, we met Olive's parents, Ernest and Vashti White.

I resigned from GE and we moved about the middle of December. We were graciously invited to stay temporarily with Olive Gambrell's parents, the Whites, until we could find a place of our own. After a week, we found a very nice duplex apartment in Crestline Heights, on Euclid Circle.

Immediately upon arrival, I went to work with Air Engineers. And from the moment we arrived, Ann and I both loved Birmingham, and have not changed our minds in the fifty years we have lived there.

We attended Canterbury Methodist Church, which at the time was a little pioneer country church, located between Homewood and Mountain Brook. Later, Canterbury and Mountain Brook Methodist Church merged and built a big complex in Mountain Brook, but still known today as Canterbury.

After six months of living on Euclid Circle, Mrs. Grubbs' home in East Lake became available. She was a good friend of Mrs. Gambrell and lived only a block away. Although it was an older home, it had a coal-fired central heating system.

During this time, I had to have an emergency appendectomy at St. Vincent's Hospital. When the officials discovered that Ann was a registered nurse, they recruited her. She worked there off and on for a year.

While we were living at Mrs. Grubbs' house, our first son, William, was born on February 13, 1950. Ann had an extremely rough pregnancy, spending approximately seven months out of the nine in the hospital. She could not keep anything on her stomach, and she was nothing more than skin and bones.

Earlier in the pregnancy, her doctor had considered aborting the baby in order to save Ann's life. Ann's mother was there and she asked about the chances of her surviving the delivery and the baby being all right. The doctor said that Ann had a 50/50 chance. So Ann and her mother talked it over and decided not to terminate the pregnancy.

I shall never forget the day that William was born. The doctor said that he was going to do everything in his power to save Ann. He did not mention the baby, apparently thinking there was little chance for him to survive. Some two hours later, as I was pacing the floor and praying, I heard a baby squalling from the delivery room.

Oddly enough, there was not another baby in the entire complex that day, so I knew that the cry had to come from our baby. The doctor later said that William was the first baby he had delivered that started screaming before he got out of the birth canal.

Shortly thereafter, when I was allowed to visit with Ann, a nurse brought in her lunch, which consisted of the most awful looking spinach greens. I hesitated to offer it to her, but after nine months of little food, she ate every bite as if it were delicious ice cream. To this day, she still loves spinach.

Back then, there was no medical insurance. Ann remained in the hospital for a week, when she and our dear little one were discharged. But there was a major problem, the hospital bill, which had to be paid before they left the hospital. Fortunately, because Ann had worked there as a nurse, the bill had been reduced by half. But I still needed to come up with $300.

I went to my little bank at Five Points South and asked for a loan, telling the loan officer why I needed it. He ungraciously declined. What was I to do? I went to my boss and dear friend, George Holmes, and explained my predicament, hoping that the company could advance the money based on my future earnings. He explained that he could not do this, but that he would personally loan me the $300, interest free. You'd better believe that George Holmes and I remained close, like father and son, for the rest of his life!

When William was only two weeks old, another engineer, Bill Mylius, and I went on a business trip to Tampa, Florida to bid on the air conditioning for the Hillsborough County Court House. We were the low bidders, but after much political haggling over the next few months, all the bids were voided.

I gained valuable experience at Air Engineers about the intricacies of bidding, engineering, and handling air conditioning jobs. The company was good to me, but I could see that the firm was growing more in the direction of being a wholesale distributor for large and small appliances, with less attention given to the construction department. So two years later, on February 1, 1951, I left the firm to go with Mason and Dulion Company.

All of the major air conditioning contracting firms in Birmingham were members of the union at that time. One of the union pipe fitters, Lou Hardy (no relation to the Hardy Corporation), with whom I had worked, had gone to work for Mason and Dulion. He was the one who told them about the young engineer at Air Engineers. I was offered a job at $110 a week, more than I was used to. I found out later that it was the same wage they were paying a journeyman plumber. Here I was, a graduate of Clemson, with all this experience, and I could not make more than a union plumber.

Yet, when I found this out later, it did not upset me. For decades, Mason and Dulion had been the leading plumbing and heating contractor, not only in Birmingham but throughout the state and northern Florida. They were just entering the air conditioning field. They had no one who knew anything about air conditioning. So I was strictly on my own. It was like throwing a child into deep water and letting him learn quickly how to swim. It was a great experience for me.

The first Christmas I was there, the firm gave me a bonus of $3,000, which, to me, was unbelievable. In later years, the bonus was increased to $5,000 or more. Mr. Dulion, the seventy-five-year-old owner, took a liking to me, and every morning about 9 A.M., he would bring in any mail that pertained to me and hang around in my office, just sitting and watching me work.

During the five years that I worked for him, our daughter, Leigh Ann, arrived on July 7, 1953 at East End Memorial Hospital. Ann had another rough pregnancy, although not nearly so bad as the first. A few days after Leigh Ann was born, Mr. Dulion came up to my office, and since I was quite busy on the drawing board, he did not say one word, but slid an envelope underneath my drawing paper. When I got to a stopping point, I opened an envelope to find a check for $750, with no explanation whatsoever. At the first opportunity, I asked him what the check was all about. He simply said, "I thought you might need it, son." If Mr. Dulion were alive today and running his business, I would probably still be working for him.

On March 28, 1955, while I was still at Mason-Dulion, another son, David, made his appearance in the Stegall family. He, too, was born at East End Memorial Hospital, and again Ann had a hard time with this pregnancy. Granny White had passed away in February, only one month before David was born.

Many of the people working at Mason and Dulion became my very best friends, including Nick Vance, Mr. Dulion's son-in-law, Bob Brand, our plumbing expert, and Kenny Walden, who was the heating and process piping expert. I had many good experiences while working there, including out of town bidding jobs. I especially enjoyed going to Pensacola, where Mr. Dulion's son was in charge, and helping estimate air conditioning jobs.

In 1955, both Mr. Dulion and his son-in-law, Nick Vance, passed away. We had no anchor for the ship. Mrs. Tate, a very fine and outstanding lady, served as our comptroller/bookkeeper, and Mr. Dulion's granddaughter, Margaret Vance, held the place together. But there were no bonuses in 1955.

For two or three years prior to that, Frank Hardy, owner and president of the Hardy Corporation, had attempted to hire me. We got to know each other through an engineer, Jim Foster, who worked for him. All three of us played golf.

After the first of the year, 1956, Brand, Walden, and I realized that our time with Mason-Dulion had come to an end. I knew that Frank Hardy's ambition was to go into the plumbing and heating business, and have a full mechanical contracting firm.

I took the lead and contacted Frank for the three of us. He was at the York factory in Pennsylvania. He was extremely interested and asked that we not make a move until he got back into town. We met, and he gave us a very attractive offer. We really hit the ground running, with several profitable jobs almost immediately.

One of the reasons that I was reluctant to leave Mason-Dulion was because of the sheet metal shop. Mr. Dulion had given me permission to start a sheet metal operation. We rented a building, bought all the machinery, and hired some of our old friends from Air Engineers to run the shop, including Bill Hester and Frank Hughes. Eventually, we got D.C. Lee, who was with Shook and Fletcher at the time, to head up the operation. It was a very successful endeavor.

Frank Hardy also had a sheet metal shop. After Bob, Kenny, and I joined the firm, the Hardy Corporation grew so much that we bought property and built a building on 12th Street, where Hardy is still located today. I became Hardy's Executive Vice President, and was offered a plan to buy controlling interest in the company from Hardy over a period of time, mostly from yearly stock options. Frank retired, but after about six months, he became restless and decided to come back to run the company. He discontinued the stock option plan for me. However, I had already observed that I would never be able to buy the company because the net worth was increasing at a rapid rate.

Earlier, Frank had wanted me to go to Decatur and start a business, which I declined. Later, he also considered opening a business either in Atlanta, Jacksonville, or Tampa. This was in February-March 1960. I investigated this over a period of several weeks, and actually spent an entire week with Ann in Tampa. But noticing the competition there, I came back to Birmingham and advised against starting a new business in Tampa.

After a few days, I explained to Frank that I had always wanted to start my own business, but I felt that Birmingham was the best place. There was plenty of business; the competition was slim, and everybody knew and respected me. That turned out to be an excellent move, although it was only the grace of God that kept me afloat during some very trying times.

Frank Hardy and I remained the best of friends for the rest of his life. During some hard spots in getting Stegall and Company started, Frank told me, "Stegall, whatever happens, you will always have a place here."

I had many pleasant experiences at Hardy. The business was just sheer fun. We grew to be one of the top mechanical contractors in the state, if not the Southeast. We also had many pleasant golf outings, particularly at Callaway Gardens. We would usually have two foursomes, with Frank Hardy, a few others from the corporation, and then we would pick up a couple of others, such as Stan Simpson and Frank Vaughn.

Kenny Walden and Bob Brand remained at Hardy Corporation until they retired. They both did quite well.

When I left the company to start my own business, I had no concrete plans. My first action was to pick the best sheet metal worker in the Birmingham area, and the best steam fitter/refrigeration man, and offer each a 20% interest in the company.

That sheet metal worker, the best in the business, was D.C. Lee. His coming into the company was perhaps the best decision that I made in my business career.

The steam fitter/refrigeration man was Arthur Thrasher, an excellent mechanic, with a good personality and a love for everyone.

Since I had to give up the company car when I left Hardy's, we had only the family car that Ann drove. So I got a ride over to the south side to Liberty Motors and bought an air-conditioned 1953 Mercury sedan for $300.

This was a bargain and it served my purposes well for some time.

To begin with, I rented a little twenty-five foot storefront building on First Avenue South, between 17th and 18th Streets. Actually, I took over the lease from my friendly competitor, Florence Heating and Air Conditioning. Ann and I spent days and nights down at this place, cleaning it and setting up a makeshift office. We had no machinery or tools. My first employee was Paul Sharbel, a young Alabama college student, who was out of school for the summer. He was a friend of Dave Upton, whom we allowed to build trampolines in the back of our building, where we would eventually put our sheet metal shop.

After a couple of weeks of trying to get organized and solicit business, a man named S.J. Slay came to see me. Slay was the comptroller and financial officer of a large wholesale hardware company, Wimberly and Thomas. Their office and warehouse was on First Avenue, a block away. W&T was the distributor for Chrysler Air Temp air conditioning products.

Slay wanted to employ my services as a consultant to help decide what to do about a small air conditioning company that owed them a lot of money, approximately $100,000. The company did residential air conditioning and appliance repairs. Because of the debt, all their outstanding stock had been turned over to W&T, who had two men trying to run it.

Since I was not yet earning any money, I agreed to spend two weeks looking into the company and advising W&T what to do. The two men in charge opened all their books, especially their operating statements, accounts receivable, accounts payable, and so forth. I talked to the people who had extended them credit, the supply houses, and to the customers, who owed them money. I also interviewed practically every one of their workmen, some ten or twelve, to determine their knowledge and grasp of the business. It was not a pretty picture.

Considering the personnel and their qualifications, the facilities, their customers, and reputation, I did not have any hope of the company ever recovering. I advised Slay to bankrupt the company as soon as possible and auction off the assets in an attempt to pay at least something to the suppliers. Of course, the biggest debt was to Wimberly and Thomas, while all the other accounts amounted to less than $50,000.

After a week or so, during which time I had booked our first little job, Arthur Thrasher had come aboard and I had borrowed $5,000 from the First National Bank (Red Terrell), using my insurance cash value as collateral. Before we had really gotten cranked up, Slay came to me again and wanted to know if I would take over the almost defunct engineering company, if they gave it to me. My immediate answer was, "No way. It's hard enough to start at ground zero, and even harder when a company is so deep in the financial hole."

For the next several days, Slay kept talking to me, and finally made a proposition. The owner would surrender any claim he had on the corporation, even though he had put $70,000 in the business. Since the company had a tax loss of about $90,000, this meant that I could make up to that amount without paying any income tax. To fulfill this claim, we would simply pay W&T the amount of money equal to that year's tax. He would then extend a $50,000 line of credit with which to buy sheet metal, equipment, and trucks. Slay would also get all the suppliers together and explain to them that the only recourse was bankruptcy. But if they would take 50 cents on the dollar owed them, then Stegall would take over the company.

Pleased, they all agreed. Then Slay agreed to loan me the money to pay off these suppliers.

These proposals really got my attention. There was an advantage to taking over a going concern, however shaky. The company already had two service trucks, a sheet metal truck, some basic sheet metal equipment, files, desks, and tables. Slay assigned all of this stock over to me. I immediately went to a lawyer and had the corporation's name changed to Stegall and Company, Inc., with 20% of the shares going to Lee, 20% to Thrasher, and 60% to me. This was in June 1960.

We immediately moved to the other building on 9th Street, between Fifth and Sixth Avenue North. D.C. Lee joined me, along with Catherine Millican, whom I had gotten to know when she worked for a general contractor, J.K. Holley.

Bill Walsh, an estimator, and Jimmy Patterson, a young apprentice refrigeration man, were the only ones I kept from the former company. We signed up with the local union for both sheet metal and pipe fitters and we had an excellent relationship with both unions. Buck Ward, a pipe fitter business agent, agreed to take Jimmy Patterson into the union.

My first contract job was to air condition an addition to the Ensley Presbyterian Church, with Barney Wilburn as general contractor. I think Wilburn was trying to help a struggling contractor get ahead. Harry Jeffcoat, consulting engineer, also helped to negotiate a couple of small jobs in the downtown area.

Daniel Construction Company had always been my favorite contractor in town to do business with, beginning with Mason and Dulion, then the Hardy Corporation. I continued my contacts there. They neither encouraged nor discouraged me, yet I had a strong feeling that I could depend on them for some business.

The first big job we tried to bid on was for the First Baptist Church in Huntsville, a giant job at the time. The air conditioning alone was in excess of $250,000. This was a bid job, and Daniel Construction Company was the low bidder for the general contract and we were the low bidder with them for the air conditioning. Sonny Inscho was the low bidder for plumbing. I shall always have a warm spot in my heart for Daniel and the people in charge: Hugh Daniel, Joe Denitci, John Gustafson, Wallace Cobb, Trenton Ellison, Gary Smith, Carl Crossman, Harvey Burch, and others. They showed a lot of faith in me, and in our little organization, by trusting us with this sizable and complicated job. It was a chilled water/hot water cooling and heating system with two centrifugal machines and a larger boiler.

The sheet metal was a huge part of the job. D.C. Lee hired his old friend, Paul Simpson, to take over the fieldwork for us. This was an excellent move, since very few people contributed more to our success than Paul.

After a year or so, the pipefitting end of the business could not keep up with the sheet metal end. We split up, and Thrasher went on to form his own service company. He apparently has done well ever since. Bill Walsh also left the company, and went into the manufacturer agents business, where he did quite well, also.

It was obvious that Stegall and Company needed a larger and better place. T.C. Brasfield Construction Company owned a separate woodworking shop, apart from their office. Mr. Brasfield, a fine old gentleman, had shown signs of wanting to get out of the general contracting business. He made us an excellent proposition to buy the woodworking shop on the corner of Fifth Avenue and 29th Street South, and he would even hold the mortgage. This was all the space we needed, and it was in an ideal location, diagonally across the street from Hinkle Supply Company, one of our biggest sources of supplies.

Even then, we outgrew this space in short order and had to make an addition to the building. In the meantime, Mr. Brasfield had taken in a partner, Miller Gorrie, and formed the company, Brasfield-Gorrie. Later, Gorrie bought out Brasfield completely. The new firm of Brasfield-Gorrie built the addition for us. I knew Gorrie from working with him at Daniel Construction Company. For many years, Brasfield-Gorrie was one of our favorite customers, up until the company became so large that it was impossible for us to follow them around to all the places they did business.

When the First Baptist Church was completed, it was obvious that the church needed a competent person to operate this rather complicated system. We helped recruit a young man, Ray Landers, and brought him in to train him. He took on the job there and was excellent. He finally left us and began his own business, which was very successful.

I could not do all the estimating and handling of the jobs. So I began looking for someone to come into the firm. I had gotten to know Harry Elliott as a result of his working for James A. Evans, the leading consultant engineer in town, and we had always hit it off quite well, probably because of a similar sense of humor. He had left Evans and was then working at Florence Heating and Air Conditioning. I twisted his arm and made a business proposition that he accepted. He was one of our key men for the rest of our business life, except for a few months when he worked as a consultant. But I persuaded him to return. He was dependable and people knew they could trust him. He is no longer living, but to this day, I give thanks for him and for Paul Simpson. Three other people are extremely special to me and meant so much to Stegall and Company: D.C. Lee, Catherine Millican, and my son, William. We had lots of good people, but these three particularly stand out.

Our business continued to mushroom. A few of the larger jobs were Automatic Electric Company of Huntsville, Dunlop Tire and Rubber Company of Huntsville, Madison County Court House, Madison County Airport, The Haley Center at Auburn, the Jefferson County Court House, and the Farley Nuclear Plant in Dothan.

One of our mistakes was to merge with a

plumbing contractor, and later with an electrical contractor. Perhaps we were too ambitious, and the mergers cut us off from some of our best friends, ones we could rely on for support and help.

One of our overly ambitious ventures was to bid and get the air conditioning and mechanical work on an underground building for AT&T in Lyons, Nebraska. We had done a similar job near Jasper, so we thought we were experts. The problem turned out to be the general contractor, who did not know how to build the structure. The general contractor was fired and since we were a sub to him, we also were fired, with someone owing us $300,000. Another contractor came on the job and we finished the job under a new contract, but still were not paid the original $300,000. After two or three years in court, we finally won a judgment, but had to accept 60 cents on the dollar to avoid more appeals.

We had lots of very good friends who stuck by us during this shaky time. I especially remember Buddy and Frank Hinkle, Sax Lawrence, Harry Brock at Central Bank, Ed Beal at USF&G, Leroy Monskey, with Engel Insurance Agency, and especially some great general contractors like Daniel, Richardson, Brasfield-Gorrie,

Pearce, DeMoss, and King. Daniel trusted us with two of the largest and best jobs we ever had, Automatic Electric and Dunlop, both in Huntsville.

It wasn't too many years before we were outgrowing our facilities, and we began buying property that surrounded us. We ended up owning the entire half block between 28th and 29th Streets and several lots across the alley on Fourth Avenue, as well as several lots across the street on Fifth Avenue. We remodeled an old building (Richardson Construction Company) on 28th Street and Fifth Avenue, which became our industrial shop and manufacturing facilities.

Shortly before the expansion, we had hired Paul Rettig, an excellent sheet metal worker who was particularly skilled in stainless steel. With his expertise, we began to get more and more stainless steel work. We finally ended up as manufacturers of commercial stainless steel kitchen hoods, which sold under the name of Sun-Air.

We did a world of other heavy industrial work of various kinds, even going as far as Tullahoma, Tennessee and Norfolk, Virginia. We were overly ambitious in this area, too.

Paul Rettig, who operated the industrial shop, had

an unorthodox style of management, running it without much communication with the rest of us. But the financial report each year showed that he was making money for the company.

About this time, we had a dispute with the sheet metal union and were unable to negotiate a reasonable contract. After much agony, friction, and heartache, and especially lost time, we decided that the best avenue was to go non-union. Most of our key people, like Paul Rettig, Paul Simpson, D.C. Lee, and others, were in the union, but they stayed with us. At the same time, we dissolved our relationship with the pipe fitters union. We had an excellent lawyer, Steizenmuller, who guided us through the legal technicalities. After this, the company became known as Stegall Metal Industries.

By 1986, my son William was president, and I was chairman. D.C. Lee and Harry Elliott were vice presidents. I had already given one-fifth of my stock to each of the four children, William, Leigh Ann, David, and Frank, with William holding all voting stock. I was sixty-five years old in December 1986, and Harry Elliott had virtually retired because of health reasons. D.C. Lee often talked about taking an early retirement.

Although we were busier than ever, the profit margin was less. With three of us ready for retirement, I recommended that it was time to sell the company.

Through the years, some of our greatest friends were consulting engineers, with whom we shared faith and complete trust. (1) James A. Evans was the first outstanding consulting engineer in Birmingham. We had an excellent relationship with him, as well as with (2) Harry Jeffcoat, Jr., another engineer with whom we had an important relationship. We did worlds of work with Jeffcoat as the engineer, and also with (3) Eddie Miller, who had the reputation of being a straight shooter and fine consultant.

Dave Upton was one of our original stockholders, a very valuable director and friend. I do not believe that I could have made it without Dave's encouragement and help during that first year.

No finer group could be found than the members of Stegall and Company. D.C. Lee, Harry Elliott, Bill Stegall, Jr., Sam "Hosie" King, Harry Jeffcoat, III, Roger Kilgore, Paul Rettig, Ray Landers, H.G. Jackson, Leonard Bass, Arthur Thrasher, Bill Walsh, Johnnie White, Harold Roberson, Jerry Robins, Paul Simpson,

Larry Tinney, and our number one backup girl, Catherine Millican, were all part of making the company a success. Each was completely capable of operating a business. In fact, many went on to establish their own companies, and all are doing well.

Roger Kilgore, who had worked with us for a number of years, and had done a fine job as estimator/job handler, bought the small non-union company, Cooling & Heating, Inc. that we had owned for a while. After a couple of propositions that we decided not to take, Roger, himself, came up with the idea of buying out our company. He had some rather wealthy investors, who were willing to put money into it.

Kilgore made us a proposition to buy the company at book value, take over all assets and debts, and pay us over a period of three years. We had one large job in progress at the time, the SouthTrust Bank downtown. After this job was completed and accounted for, our net worth turned out to be a little less than expected.

But we had some wonderful experiences during the twenty-six years of Stegall and Company. Each Christmas we had a special celebration with all our employees and families. We also had a company picnic each summer, with activities planned by the employees and food furnished by the company. It was always a special time for the children.

And the adults had good, clean, wholesome fun, too. Our key people, headed up by D.C. Lee, Paul Simpson, Paul Rettig, Harold Roberson, and others, went on an annual fishing trip of three or four days to the Chandeleur Islands to catch speckled trout and redfish.

There were tragedies, also, within the company. In the 1960s, when Ann and I were returning from a free trip given to us by our air conditioning manufacturers, Dave and Ann Upton met us at the Birmingham airport. Dave said that he had bad news, that some of the boys had been hurt. At first, we thought it had to do with our own boys. There was an element of relief, mixed with guilt, when we discovered our boys were safe. But it was still heartrending to find out that Andy Bearden and Harold Roberson had been severely burned in the sheet metal shop. One was doing electric welding and it grounded through a large can of glue thinner. The can exploded and set fire to everything around it. Critically injured, both men lingered for weeks. We finally lost Andy. Harold Roberson survived and is doing well to this day.

The final papers for the sale of Stegall and Company, Inc. were signed on December 31, 1986, and handled by attorney-at-law, J.N. Holt.

Roger, the new owner, had enlisted ten investors in Birmingham, and together they put in $600,000, which was used to pay off the company's debt at the bank. Then, he cut the volume of business in half, concentrating on the manufacturing end of the business, and he let half the employees go. After three years, each one of his limited partners made back his money, and the company was sold to Morris Hackney Enterprises for three million dollars.

The company has seemed to do well through the years, with the one company divided into three. Sun-Air, the manufacturing company, was sold. The special fabricating division continues to be called Stegall Metal Industries, Inc. The contracting end of the business is now known as Stegall Mechanical, Inc.

The company continues to have a good reputation, although it is not considered one of the really big companies. But for a number of years, much of the good air conditioning work in Birmingham and north Alabama was limited to three companies—Shook and Fletcher, Hardy Corporation, and Stegall and Company.

Chapter 19

An Unforgettable Business Friend

During the years that I worked for Mason and Dulion, the Hardy Corporation, and Stegall and Company, we did business occasionally with a general contractor, Barge-Thompson, Inc. from Atlanta.

Over a period of many years, this company did all of the Southern Bell telephone work here in the Southeast. The architectural firm involved was Warren, Knight, and Davis, located in Birmingham. John Davis, Sr. was the supervisor.

The company that I worked for at the time was seldom the first choice for the mechanical work and air conditioning, but we seemed to do enough work with Barge-Thompson to have at least one job going. Since I spent much time supervising the jobs we did, I got to know rather well the old gentleman, John Davis, the architect, and Barge-Thompson's project manager, Bill Gifford. For some reason, Bill and I hit it off quite well. Part of this might have been that we were both from South Carolina.

Bill Gifford had been born on a huge plantation in South Carolina, just across the river from Savannah, Georgia. A little country town, Gifford, is still in existence. It was apparently the control center for the plantation.

As to formal education, I'm not sure about his credentials. If he were not a university man, then Bill had dug out an excellent education on his own. I have never known a more intelligent person. He could converse knowledgeably on almost any subject that came up.

While he was a young teenager growing up on the plantation, he loved to go quail hunting on a mule. One day as he got near a covey of quail, the hunting dog suddenly flushed the birds. The whirring sound and the movement caused the mule to buck just as Bill raised his shotgun to shoot. The reaction of the mule threw his aim awry, with the gunshot tearing off several fingers on one hand.

The mule threw the rider off and ran back to the barn, which was some distance away. Bleeding profusely, Bill managed to walk or crawl the distance home. He survived, although he was left with only two fingers on one hand.

Several years prior to World War II, a plane made a forced landing on a beach somewhere in the vicinity of Hilton Head Island. It was severely damaged and was abandoned, with no one claiming the wreckage. Bill and his father, after a suitable time, brought the plane back to the plantation.

Since the two were flying enthusiasts, they took great pleasure in rebuilding the plane and teaching themselves to fly. By the time World War II started in Europe, Bill had secured his pilot's license. He tried to volunteer for the U.S. Army Air Corps, but because of his missing fingers, he could not pass the physical. That did not stop him. He simply went to England and joined the Royal Air Force. There he was trained to be a fighter pilot. He flew a Spitfire during the siege of London, and was successful in many other missions.

Shortly after Pearl Harbor, on December 7, 1941, when the Japanese destroyed the U.S. fleet based there, the U.S. Army Air Corps called on all Americans who were flying with the RAF to transfer to the U.S. Army Air Corps. Bill took this opportunity to come back home and fight for his own country.

In the United States, Bill was sent back to school to learn to fly the gigantic B-29 bomber. This airplane gained fame flying out of Saipan and hitting the mainland of Japan. Bill was one of the pilots.

Flying with him on one mission was a very special rider, Ernie Pyle, who was one of the greatest writers to cover World War II. He was especially good at writing from the infantryman's viewpoint. Ernie Pyle said of Gifford that he was the only man who could fly a B-29 with one hand and hold a glass of bourbon in the other.

After the war, Bill came home and began his business career with Barge-Thompson. On our visits to the telephone job sites, particularly on final inspection, a whole group of us would meet and celebrate at the best eating place in town. This group included Bill Gifford, John Davis, who usually picked up the tab, and a few of the telephone dignitaries. Since I was the only one in the group who did not drink, I had a hard time persuading these folks at the end of the evening to leave the restaurant and go home, since the next day was also a work day.

I recall on one occasion that Mr. James A. Evans, the consulting engineer who always designed the mechanical work, was a member of the group. That evening, Davis and Gifford began discussing the Dead Sea Scrolls, which had been found. They both had had a few drinks, but that did not stop them from having a very lively and intelligent conversation.

In Bill's latter years, he would work at the Barge-Thompson office all day and get so absorbed in his work that he would not realize that everyone else had gone home and that it was eight o'clock in the evening. He would then stop off at the package store, and once home, he would continue his study of the Dead Sea Scrolls.

Gifford never weighed more than one hundred and twenty pounds, but he was able to attract an unusually intelligent lady to marry. Their intelligence was passed on to their son, who was so superior to the average school age kid that he went to a special school for the extremely talented.

One evening, Gifford brought along his son to the restaurant. He was about eight or ten years old at the time, and he, like his father, was able to talk about any subject brought up in discussion.

Although his drinking never got out of hand, I believe his unusual lifestyle may have contributed to his short life. He died when he was in his late fifties.

But one thing I do know. Bill Gifford was one of the most talented and unforgettable business friends that I ever had.

Family Vacation,1965
(L to R) Ann, David, Leigh Ann, Frank, and William

The Stegall Family
Stegall, William, Ann, Frank, Leigh Ann, and David

Chapter 20

EUROPEAN DIARY, 1965

After finishing Clemson College and beginning work, I had never taken time off for more than a week. In 1965, Ann and I decided to travel to Europe on a three-week family vacation with our four children—William, age 15; Leigh Ann, age 12; David, age 10; and Frank, age 8.

Since this was really the first vacation of any kind that we had taken since I started Stegall and Company in June 1960, it was going to be something special. AAA was most helpful in furnishing maps and tour books. I also ordered from Michelin detailed maps of certain sections in Europe, where I had served in the 121st Cavalry Reconnaissance Squadron. For several months, we dreamed; studied these maps; and made plans.

To help finance the trip, Ann had saved money for twenty years. The bulk of the savings came from my monthly disability check that I received as a result of my war wounds in World War II. To that small stipend, Ann had added any income tax refund and any other unexpected money that had come our way. The total had grown to $3,000.

The round trip airline tickets from New York to London cost $300 for adults (William was considered an adult), and $150 for children. We flew in a propeller plane from Birmingham to New York at the night rate.

On Tuesday, June 1, I worked furiously all day, seeing to the paper work and signing checks at the office, in preparation of my being away for the next three weeks. Our business was going well at this time under the able supervision of Harry Elliott, D.C. Lee, and others.

About seven P.M., I arrived home in Vestavia, the residential community where we lived, and the family had dinner at Britlin's. Bonnie, William's girlfriend, and Diane Brown from next door came by for a farewell party, and after that, we finished packing. When we got to the airport to catch the 1:42 A.M. plane, we were pleasantly surprised to find that our friends, Harry and Martha Jeffcoat, were waiting to see us off. Here it was, in the middle of the night and in the middle of the week, and the two had come to wish us a pleasant journey.

After a stop in Atlanta, we flew on to New York, arriving at Kennedy Airport at 7 A.M. Since the plane to London would not leave until that evening, we checked into the International Hotel. We were given a special rate of $11.65, based on our vacating the room before 6 P.M. We all lay down and slept for several hours.

The children had never been to New York, so we all went downtown to see Rockefeller Center, the United Nations Building, the Statue of Liberty, and other sights. We took a bus from the hotel to Key Gardens and then the subway to 50th Street downtown. On the way back, we returned by subway from South Ferry, changed at Central Station, again at 74th Street, then on to Key Gardens, where we took a bus to the hotel. Our total expenditure of that outing, not including the hotel, was seventeen dollars.

At 7 P.M., on June 2, we departed Kennedy Airport in a Boeing 707. Actual flying time was five and a half hours. It was already morning in London, when we arrived on June 3. We went through customs and then took a bus to Victoria Station.

Through AAA, we had made arrangements to buy a Volkswagen station wagon, to be picked up in London. At that time, the model was not available in the United States. It had all the appearance of an American station wagon, except for the square back.

As soon as we arrived at Victoria Station, I called the dealer. Since the agency did not open until 9:30 A.M., we checked our baggage and walked down the street to the Grosvenor Hotel, where we ate breakfast.

The dealership was located in the north of London, so William and I rode a double decker bus across town. When we arrived, we discovered that the VW was not there. It had to be brought from a distance away that required another two hours.

The VW dealer was very nice, however; and sent a car to pick up Ann and the children, inviting William to go along. He even offered to provide us a hotel room to rest while the paper work was done. Hoping it would not take much longer, we declined. After paying the delivery charge of $56, we left the dealership in our new vehicle at 11:30 A.M.

We drove from one end of London to the other, from the northwest section, through Westminster, by the House of Parliament, to the southeast section. The car was new and strange, with a manual shift. And we were driving on the wrong side of the road, through the worst traffic of London.

The Lord had his angel watching over us.

While driving on the freeway, we all became quite sleepy, including the driver. So we stopped at the first place we could find for coffee. At Canterbury, we bought gasoline, which cost four times as much as it did in the States. We had the idea of taking time to visit Canterbury Cathedral, but as I talked with the service station attendant about plans to go on across to France, he gave us some firm advice. It was a holiday—Ascension Day, which always fell on a Thursday, forty days after Easter. He told us that we should go immediately to Dover, if we wanted to get across.

Normally a person would buy a ticket prior to getting on the ferry. But since it was leaving, the officer motioned us on and he took the fee, $47 for the car and its occupants. We were the last car to drive onto the ferry. Once again, the angel was with us.

After a nice hour-and-a-half trip, the ferry docked at Boulogne, France. It was a beautiful harbor and very picturesque. It had been our understanding that the ferry was going to Calais. But we could see from the map that Boulogne was just a few miles down the coast, southwest from Calais, so it did not disturb us.

When we had first planned our trip, we decided not to be tied down to specific reservations. It was now about 9 P.M., so we looked in our *Europe on $5 a Day* for an inexpensive hotel or bed and breakfast. We chose the Marmin Hotel and although the people directing us did not speak English, we understood the general directions.

When we reached the hotel and checked in, we found that the dining room was already closed. It was reopened to serve us dinner. We all slept soundly that night. When we went down to breakfast the next morning (June 4), we discovered that we were actually in the Excelsior Hotel, another small hotel in the same building. But no harm was done. The total cost for the rooms, dinner, and continental breakfast was $28.50.

We were a religious family, with just a general idea of the places and countries that we wanted to visit. So it seemed natural to ask the Lord's guidance each morning at devotions. We took turns in Bible readings and prayers, and on that particular morning the fellowship seemed especially meaningful.

After breakfast, we left Boulogne around 10:30, and drove through the beautiful countryside and along the seashore, going from Calais, and Dunkirk, and on to LaPonne. In one small town we bought fruit and ice cream. Frank had become sick and he was no better that day. We spent considerable time walking through the town in search of a doctor. Although we did not find one, the walk seemed to help Frank, as well as the soup he and the rest of us ate for lunch in a very nice family restaurant. It was in this town that I got a good shave for fifty cents.

At Oostende, Belgium, we left the shore drive and got onto the Autobahn. It was a beautiful drive to Brussels. After that, we had planned to visit the old battlefield of Waterloo, where Napoleon was finally defeated. However, we missed the turnoff, so we decided to keep driving and go down to Bastogne, the city famous for the Battle of the Bulge at Christmastime, 1945.

After a nice visit at Bastogne, we drove on into Luxembourg, a very unusual and charming city. We had picked out the Molitor Hotel, and we found it without any problem. That evening, we ate dinner at the Wallsheim Restaurant for $14, including tip. By midnight, we were all in bed and enjoying the eiderdowns.

On Saturday, June 5, we got up at 8 A.M. and had our continental breakfast. The hotel cost for all of us was $16.50. That morning, the rain and mist made it difficult to take pictures, but we were extremely impressed with the large steel bridge that crossed the gorge through the heart of Luxembourg.

From Luxembourg, we went on to the well-kept and beautiful U.S. Military Cemetery at St. Avold. I explained to the kids that I would have been buried there if I had been killed in the mine explosion in 1945.

We then drove through various small towns until we reached Ludweiler, Germany. By that time, the rain and fog had increased. But I was able to find the house where we had our platoon headquarters for the 121st Cavalry, and the outpost where our armored car had dug in. That area was fenced off and the road closed.

To Saarbrucken and up the Saar River, we traveled to Eyweiler, where I had spent the night before the explosion. We did not stop, but continued through other country towns to Strasbourg, France. There, we located the Autobahn. Crossing the Rhine River, we arrived at 8:30 P.M. in Karlsruhe, Germany. Mike Welden, Ann's nephew, was stationed there. He met us and then traveled with us for the next week or so.

We stayed in Karlsruhe two nights—Saturday and Sunday, June 5 and 6. Here, we were able to wash our dirty clothes, go to church services, and explore the city. We secured a good map and, using it, we found the zoo, about ten blocks from the hotel. Everything in the city was so clean, and the German families were all dressed nicely. The stores carried very interesting articles of high quality. Since I needed to change some dollars into deutschmarks, Mike, William, David, and I went to Mike's army base. When we returned, Ann's leg was swollen. With special care, it did not become a serious problem.

On Monday, June 7, we left Karlsruhe for Zurich, Switzerland. On the way, we went through back roads and visited some of the places I had seen in World War II. As we passed by Merlebach, along the French-German border, where our army unit had rested for several days, I remembered especially the hot showers we had taken at the coal mining facility.

We also went through Forbach, where our unit

had dug in, expecting a German counterattack that never materialized.

William did much of the driving that day. Taking the Autobahn, we entered Switzerland at Basel, and by 8:30 P.M. we arrived in Zurich. The hotel that we had selected was full. A very nice lady called and found rooms for us at the Bristol Hotel. We finally had dinner at 11 P.M. at the train station, but the meal was very good.

We slept late, until 9 A.M., and then enjoyed an excellent continental breakfast. I drove the VW to a dealer for warranty service, which was going to take all day. So Ann and Leigh went to the hairdresser, while the boys decided to go to Globus Department Store. We met again at lunchtime.

We had learned that it was standard procedure for Europeans traveling to have a picnic at lunch. That seemed a good idea, and it was also cheaper. We bought bread, meat, strawberries, apples, and cheese for our lunch that day.

Close by our hotel was a large indoor swimming pool, 50 x 15 meters. William, David, and Frank wanted to go swimming, which they did. But they soon discovered that the water was extremely cold. After their swim, the family went to Jelmoli Department Store, where we shopped until 5 P.M. For $68, I bought a very nice camera, which I used for a number of years, before finally giving it away. The entire family really enjoyed that full day of shopping, swimming, and eating good food.

On Wednesday, June 9, we ate a late breakfast and went on a tour of Zurich. We rode cable cars to the top of a mountain, where we had a beautiful view of the Swiss countryside. We went shopping again and then had a leisurely drive over to Lucerne. At 4 P.M., we checked into Motel Lucerne-Sude, where we rested and again washed clothes. William and Mike decided to sleep, so they did not go with the rest of us to dinner at the Waldstatterhol, where we ate the local lake fish. Someone said that there is no such thing as a bad hotel or restaurant in Switzerland. I agree. This was our best accommodation yet ($22.50 for all seven of us), although Germany also ranked high.

On Thursday, June 10, we left Lucerne on our way to Lugano, Italy. The weather had turned cold and it rained all day. Because of the weather we did not go sightseeing. Instead, we shopped. Ann bought several watches, and I bought three rods and reels and fish knives for my friends, D.C. and Paul, and for me.

We drove around the beautiful lake and into the mountains. Again, we bought groceries for another picnic on the road. We had filled the car with gas before leaving town, but the transmission in the VW was not acting correctly. In Andermatt, near St. Gotthard Pass, we found a VW dealership. Famous as a winter ski resort, the town had the only auto service place for hundreds of miles. Once again, our angel was watching over us. The repairs on the car were badly needed, but they were covered by the warranty. After two hours, we were on our way again and headed toward Lugano.

Since we expected to arrive late, we called ahead to the Vezia Motel, so that rooms would be held for us ($21.50 for all). With snow, glaciers, and lakes, the scenery around Lugano must be the most beautiful in the entire world. That evening, we ate dinner in the car, since we had bought plenty of groceries in Switzerland. By 10:30, we were all in bed.

Leaving Lugano at 9:30 A.M., we stopped at a VW place to have the car tank filled with gas and the heater cut off. The weather was somewhat overcast, but it had turned hot. We drove to Guiness, and at a miniature village, Melive, we crossed the Italian border and changed dollars into lire, with one dollar equaling 620 lire.

Very soon we hit the autostrada, or expressway, which was a beautiful drive. William did much of the driving again on the expressway, going 60-80 mph. When we stopped for our lunch along the road, we met a McCall family from Montgomery, Alabama.

By 4 P.M., we arrived at the entrance to Venice, and found rooms at Hotel Cavaletto Doge Orseolo. We parked the car on the top of the parking garage and locked it. Most of our bags went to the hotel by motor boat. The porter from the garage went all the way to the hotel. His tip was 1,500 lire, with the boat ride costing 7,000 lire. At the hotel, we had two rooms for the seven of us, but no private bath. It was down the hall. Otherwise, the hotel was very nice. The price of the room included dinner and breakfast, and the children were impressed with the many courses served at dinner.

We walked around sightseeing and window shopping until 11 P.M. We had never seen a place like Venice, with children feeding the pigeons in the Piazza San Marco, or St. Mark's Square. We made a tour of the Basilica San Marco, where tradition has it that St. Mark is buried. We were not the only Americans in Venice. We saw many more, especially Yankees.

On Saturday, June 12, we slept late. After getting dressed for the day, we had our usual continental breakfast, except for David, Frank, and William, who ordered a special breakfast.

We all went on a tour through Venice and traveled across the bay to Murano, to a glass works factory, where Ann bought a lot of crystal, using travelers checks. By the time we went to Lace Place, we were using our American Express card.

For 6,000 lire, we all went on a one and one-half hour gondola ride, through many small canals and twice through the Grand Canal. I was expecting a very romantic ride, but evidently we chose the wrong time of day. People's laundry hung out the windows, and garbage was being dumped into the canal.

We also climbed to the top of the bell tower that day, where the bells were always rung at 2 P.M.

Our children loved the delicious Italian ice cream, and had some several times a day. They also wanted to feed the pigeons in St. Mark's Square, so for 15 lire, we bought a small bag of corn, which the pigeons loved. Ann and I enjoyed tea in the courtyard and listened to one of the four orchestras, while the children were busy feeding the pigeons.

Later, I went to the room, rested, wrote several post cards, and brought my diary up to date. By 7 P.M., we were all bathed and dressed for dinner. The soup was excellent, as well as the main course, with dessert and fruit. After dinner, we went for a walk along the Grand Canal. Our hotel faced the west side of the lagoon, and was the center of much activity. The noise and singing lasted until late at night, although we were in bed by 11.

The next day was a Sunday—June 13. After our breakfast and devotions, the three younger children wanted to feed the pigeons one last time. At 10 A.M., we left the hotel by motor boat for the parking garage. We loaded the car, filled up with gas, and by 11:30, we were headed for Verona.

We did not know at that time, that Venice would suffer a devastating flood a year later, with so many of the sites that we had seen, damaged by the water. The restoration is still ongoing.

Leaving Venice, we made good time, especially on the freeway. We stopped once for ice cream and fruit. The drive, alongside a river, was beautiful. About 4 P.M., we crossed the Italian-Austrian border, where we changed some American dollars and our remaining lire for Austrian schillings.

At the famous Brenner Pass, we crossed the Alps and went on to Innsbruck, taking the old road. It was some of the most beautiful scenery we had seen. Since the pension we chose from our guidebook was full, we drove around until we came to Salzburg Road. We just happened to stop at a very good accommodation. We ate dinner at Olympic Village, a beautiful view. Everyone seemed in good shape, and we were in bed by ten.

The weather on Monday, June 14, was somewhat cloudy and the temperature ranged from cool to mild. Our total cost for rooms with breakfast came to $14. Before 10 A.M. we left the pension and visited Olympic Stadium. We rode up the mountain to the bobsled course, where the children walked around to the top. Ann elected to remain in the car. I walked partway with Leigh, then bought drinks for the children and talked with the people there. Back and forth and up and down the mountains we rode, finding the way to the ski jump, stopping at a grocery store for food, and eating lunch at a picnic area. When we took the cable car to the top of the mountain, which is over 2,300 meters high, Ann chose to remain below.

That afternoon, we drove on to Oberammergau via Garmish. It was very mountainous and the car had to use low gear often. We crossed the Bavarian border at 4 P.M., changed our money back into marks, and looked for a place to stay. We got three rooms and a great dinner for $15 at "Schnitzerstube," a pension owned and operated by a Mrs. Rutz. Before World War II, Mrs. Rutz had spent a year in London, so she could speak and understand English quite well. She welcomed us graciously, and she did many things to make our visit in this little city a wonderful experience. She had an eighteen-year-old daughter, Krystal, living with her at the time. Although Krystal could not speak English as well as her mother, she understood the language, so I enjoyed talking with both Mrs. Rutz and her daughter.

On Tuesday, Mrs. Rutz looked after the little ones while Ann had her hair done and then we went shopping. Many places sold woodcarvings. After another picnic lunch, which we ate along the way, the family visited Linderhof Castle. In designing the castle and grounds, which were quite unusual, the architect had evidently gone rather wild, in my estimation, using everything except brains. We spent a great deal of time touring the grounds, and the we left about 2:30. It took us two hours, driving north to Augsburg, on a road still under construction. By 4:30, we were on the Autobahn, and we began to make good time. Again, William drove much of the way and did well.

It was still daylight when we reached Karlsruhe at 7:30 that evening. After spending ten days with us, we dropped Mike off there. His buddies had just returned from maneuvers, so it was a good time for Mike to get back, also.

It was a long day. We drove on to Mannheim, where we had dinner, and then on to Worms, where we found three rooms at the Dom Hotel. By Wednesday, June 16, we got up early to begin our sightseeing. This was the city where Martin Luther had appeared before the Imperial Diet in 1521 and had been ordered to recant his beliefs. When he refused, that began the Protestant Reformation. We saw his monument and visited Old Cathedral and Dom Cathedral. I could have stayed all day at a very historical old museum, that told the story of Worms and the Rhine Valley from 54 A.D., but by lunchtime, we were pressing on to Frankfurt. The traffic on the Autobahn was very heavy, but by 2 P.M., we found ourselves at the exit to the Frankfurt Airport.

Someone had advised us that, if at all possible, we should visit Berlin. By now, we were running short of time and had just about given up the idea of going. But we decided to exit and find out if there might be an inexpensive flight to Berlin. We drove into the airport parking at ground level, and went inside.

Pan Am had a special round trip price of $25 for Americans to fly to Berlin. Since the plane was leaving at 2:30, we hurriedly packed our toilet articles and pajamas in one suitcase, and boarded an old DC-3, a propeller plane that flew at 3,000 feet. At that altitude it gave us an excellent view of the countryside, which was primarily in Russian hands.

A little before 4 P.M., we were in Berlin. We immediately went to the tourist information desk to inquire about rooms for the night. The desk clerk made reservations for us at the Vogata pension.

When we arrived, we discovered that there was a different pension or hotel for each floor. Our rooms were on the fourth. For six of us, the price for one night was 65 marks, which came to about $15. After we had rested a little while, we walked down to the center of the city. It was the most up to date place we had seen, with many shops, excellent merchandise, and well-dressed people.

We stayed for a long time at an outdoor café, with the children eating hot dogs and ice cream. Frank, the ice cream lover, especially enjoyed his particular flavor. By the time we returned from the center of Europa, where there was an ice skating rink, the kids were tired, and went on to bed. Ann and I ate dinner at an excellent restaurant in the hotel. The price for two was $4.50.

June 17 was a holiday in West Berlin. The celebration had something to do with a revolt in East Berlin in 1953. We got up early, ate breakfast, and checked out. Before leaving the hotel, we bought sightseeing tickets.

Carrying our luggage and purchases, we walked three blocks to catch the tour bus. We began our tour about 10 A.M., and saw many historic spots, including the Reichstag, the former Nazi headquarters.

There were many beautiful apartment houses in West Berlin. We saw Olympic Stadium that seated 100,000 people during the 1936 Olympics, as well as the smaller swimming pool stadium that seated 8,000.

By afternoon, the tour bus took us into East Berlin. That portion of the city seemed quite drab, unkempt, and not nearly so prosperous. Most of the buildings were still in ruins, and those in use were not painted.

Going through Checkpoint Charlie from west to east had been quite an ordeal. And it was even more so when we returned to the west. The Russian soldiers checked off each name on their list to make sure that no one was trying to slip through. They checked off all our names, except Ann's. Ann almost got to the panic stage, thinking that she would be left in East Berlin. But they finally found her name on the top of the next page. It had been covered over by the fastener on the clipboard.

With the tour over at 2 P.M., we caught taxis back to Tempelhof Airport. The timing was excellent. We ate a snack for lunch and then boarded the 3 P.M. flight for Frankfurt. That morning, it had rained, and then turned cloudy. It was raining again as we left Berlin.

We arrived back at our car in Frankfurt at 4 P.M. We drove on to Wiesbaden and on down the east bank of the Rhine River, stopping off at Rudesheim. The gift shops and the ice cream places were open, and I enjoyed chatting with some of the people who spoke English.

After another beautiful drive down river, we stopped about 9 P.M. in Koblenz, a veritable storybook place. Staying overnight at the hotel, Sophienhohe, in Koblenz-Tfaffendorf, we ate dinner that night at the hotel, which overlooked the Rhine.

The next morning, June 18, the weather was overcast, with wind and rain, but as the day went along, the weather improved. When we left the hotel, Ann gave the proprietress a Kennedy silver half-dollar, which she much appreciated.

We crossed the Rhine and drove up the west side toward Bonn. On the way, we stopped off at a small town south of the Remagen bridge and bought groceries, stamps, and changed currency. We took a picture at the bridge site. Before we got to Bonn, we entered the Autobahn and stayed on it until we got off in Holland.

We crossed the border, went through customs, and again changed currency. We went on to Laren in Utrecht rather than Laren in Gelderland, where our friend, Cornelius Vanderbragen, known as "Cornie," lived and had his missionary headquarters. Since we were in the wrong Laren, and it was late, we decided to drive on to Amsterdam. It was rush hour, with cars and bicycles going in all directions, with seemingly little traffic control. We finally got through the city and bought gas. Seeing an American-looking hotel, we stopped. We got two nice rooms, each one having three beds, heat, and a private bath. While I rested, Ann washed clothes, and our kids played with some U.S. Air Force kids from Maine.

We ate dinner in the hotel dining room, with live piano music in the background for the guests' enjoyment. We had the kids to go up and ask the piano player if he knew "Georgia on my Mind." I think he was American, and he immediately started playing it. The hotel was close to the airport, and later, when Ann and I had coffee, we talked with a pilot from Tampa, Florida.

It took a long time the next morning, Saturday, June 19, to get assembled, pack, and leave. I plotted a route on the map to get downtown. We parked at Dam Square, and went shopping. The city seemed much friendlier, with a little less traffic. William bought a ring; David and Frank bought wooden shoes. I bought Mama a watch. Everyone enjoyed shopping, especially after splitting up. We shopped until noon and then began a tour north. We drove by a large canal connecting Amsterdam to the North Sea. The Grand Dike was 26 kilometers long. Just before crossing the dike, we bought groceries, stopped at a monument, ate lunch, and took pictures.

We had almost given up hope of finding the right Laren, but as we drove south on the east side of the lake, Ijsselmeer, we found the exit. We arrived at the missionary outpost at 6:15 P.M. Cornie was the first person we saw. He was in the flower garden, working with a crew of assistants—Millie and Charlotte from Pennsylvania; Helen from Connecticut; Maraja and Nellie from Holland; and a Dutchman, Braum, his right-hand man.

A few days prior to our European trip, Ann and I had been to a wedding party at the Jeffcoats', whose son, Harry III was marrying Katherine Freeman from Thomaston, Georgia. When the mother of the bride discovered that one of the countries we planned to visit was the Netherlands, she told us about Cornie Vanderbragen, and was enthusiastic about our going by to see him at the Bible missionary outpost that he ran. At that time, we had no idea that we would be in the vicinity.

We immediately fell in love with Cornie and his helpers in the ministry. He is an excellent photographer, a writer of many books and articles, an evangelist, and a teacher. We wound up staying two days with them, although we reserved three rooms at the Locham Hotel.

We spent two hours with Cornie that night, and the next morning, as we were getting ready to leave, he called and invited us on a picnic to the only mountain in the Netherlands. It was one hundred feet high and near the German border. We stayed at the hotel and waited for Cornie to arrive by bus. He appeared by noon, and we went for a ride to see old castles and interesting places, such as the Canadian cemetery. Three of the ladies, Millie, Nellie, and Maraja, rode the ten miles to the picnic site on their bicycles.

After the picnic lunch, Cornie conducted a Bible class for us. As we participated, it all seemed so easy and natural. Leigh Ann attributes that experience as a major step in her religious life.

Cornie and his assistant, Braum, have visited us many times in the States, and we have been back to the DeVluchtheuvel at least once. Ann spent two weeks there at a Bible conference. Through the years, we have tried to lend a little support to Cornie and his mission. Nowadays, he spends half his time in Miracle Manor atop a mountain in Hiawassee, Georgia.

After our mountain outing, we returned to our

hotel a little after 5 P.M. We rode on to Königen, where we spent the night. The next day, June 21, we had a good Dutch breakfast, and planned to drive on through to Dunkirk, to catch the ferry. The hotel manager told us that the nearby ferry was running three hours behind, and another was out of order. We decided to go to Dunkirk by way of Antwerp and Gent, Belgium. We were very fortunate to get through the tunnel at Antwerp because of all the huge trucks. By the time we got to Dunkirk, the 2 P.M. ferry had just left, and another would not leave until 6 P.M. So we drove on to Calais, and caught a ferry at 4:45 P.M.

This was a good ocean-going ship, with dining room. We ate dinner in the dining room, and arrived in Dover about 6:30 P.M. After going through customs and all the red tape, we drove on to Canterbury, switching once again to the left side of the road. Filling up with gas at the Shell station, we drove on to London. Getting directions from a bobby, a British policeman, we crossed Westminster Bridge and went on to the Strand Palace Hotel. We were very tired after the long trip, but this hotel had no vacant rooms and no parking. The hotel called another, the Regent Palace at Piccadilly Circus, and reserved three rooms for us. We parked in a three-story garage nearby. Accommodations were excellent, and we went to bed as soon as possible, about 11 P.M.

On the morning of June 22, we awoke to typical London weather—overcast and rainy. We had an early breakfast and hoped to take an all day tour of the city, but that tour was filled. So we made reservations for a shorter one that began in the early afternoon and lasted until 6 P.M. That left the morning free.

William and I went to the AAA office to ask about turning over our car for shipping back to the States. We then drove the car to the shipping agent near London Bridge and made all the necessary arrangements. We rode the subway back to the hotel.

Ann had been busy, confirming our flight reservations and writing cards. We all assembled and took the subway to Russell Square, where the tour would begin at 1:30. The tour was interesting and included Eton, Windsor, and Hampton Court. We spent an hour at Windsor Castle, mainly in the chapel. We spent a half-hour at Hampton Court, and took two rolls of film of the beautiful flowers, but the film was lost somewhere along the way home.

When we completed the tour, we took the subway to Tower Bridge, walked across, and continued up the other side of the Thames. Then, we rode the bus back to the hotel. That night, we ate dinner in the Minuet Room. All took baths, read awhile, and then went to bed.

On Wednesday, June 23, we had a substantial English breakfast. This was our last day in London, so we took the subway to Buckingham Palace to watch the Changing of the Guard. We had retained one room at the hotel until it was time to leave for the airport. After walking back to the hotel, the children stayed in the room while Ann and I did some last minute shopping. We also bought an extra bag to carry all the purchases we had made, including earrings, charms, and bracelets. Getting back about 1 P.M., we finished packing while the children went shopping. We gave each a little over a pound, and they enjoyed selecting their purchases.

When it was time to leave the hotel, we took down our seven pieces of luggage and flagged a taxi. We had planned to ride in the taxi to the train station, but the driver suggested that it would be cheaper for him to take us on to the airport. That was probably true. It cost us six dollars, and the taxi ride was much more enjoyable.

Checking in with Pan Am at 3 P.M., we had a wait of three hours. That gave us an opportunity for some last minute shopping in the duty free shops, after we had passed through customs. We spent all of our English pounds, and boarded the Boeing 707 shortly after 6 P.M. Due to arrive in New York at 9:05 P.M. (we lost a number of hours going home), we flew at 35,000 feet via the northern route. One hour after take off, we had an excellent meal. Although it was a smooth flight, William woke up feeling ill, and he lost his dinner. He later ate a second meal.

Thursday, June 24 arrived while we were in flight. David and Frank played with a royal horse and soldier set, while I read and wrote in my diary. About one hour from New York, it grew dark. We landed promptly at 9:10 P.M. and we went through customs without incidence. One of our bags had a broken zipper, so we put its contents into another bag. Pan Am handled getting the bags to Eastern, for our flight back to Birmingham. We took the bus to the Eastern terminal, and waited to board the plane scheduled to leave for Atlanta.

We had two hours of sleep and then woke up cold and rather stiff after such a long journey. We arrived in

Atlanta at 1:00 A.M. We had an hour-and-a-half wait in Atlanta for the propeller plane, and we finally arrived in Birmingham at 4 A.M. A car had been left for us in the parking lot, so we loaded all our bags and drove home, after being away for a little over three weeks.

While we were crossing the ocean, the six of us voted on our favorite countries. We had a great time in all, but the English-speaking country won out because we were able to communicate. When all the votes were cast, England came in first, followed by Switzerland, the Netherlands, Germany, Austria, Italy, Belgium, Luxembourg, and France.

We have been on a number of interesting trips since 1965, but this first one with the entire family was the best vacation ever.

Stegall Famly in Holland, 1965
Ann, Frank, Cornelius Vanderbragen, William, Leigh Ann, David, and workers with Reaper Fellowship

Chapter 21

Return to Europe

Since our vacation trip to Europe in 1965 had been such a success, Ann and I determined to take at least one nice trip each year.

But the next year, we decided that we would like to see America. So we drove our Pontiac station wagon with all four children all the way to California and back, with stops along the way.

One morning, we left San Francisco and drove down the coast road toward Los Angeles. We saw the Monterrey Peninsula, Pebble Beach, and some of the most beautiful scenery ever. We took our time, driving quite a distance, and finally stopped for the night.

We were late getting up the next morning. As

William and I were packing the car, he saw three girls across the parking lot, and he called my attention to one of the girls. "Doesn't she look exactly like Krystal Rutz, the girl we met in Oberammergau last year?" I answered, "Yeah, she sure does." And I went on with the packing.

About 11:00 A.M. we decided to stop for a late breakfast or early lunch. We found a suitable looking place, so we went inside. As we were getting ready to order, William said, "Look! Over there are those same three girls we saw in the parking lot, and one of them still looks like Krystal Rutz from Germany."

We dared each other to go over to their table, but no one took the dare. But as we got up to leave, Ann went over and said, "Aren't you Krystal Rutz from Oberammergau?" She looked absolutely startled. "And you are the Stegalls from Alabama." One of the other girls, who was an American, said, "No one will ever believe this."

Two or three years later, our firm was doing an air conditioning job near Omaha, Nebraska, and I was flying back and forth from Birmingham to Omaha. On one trip, I picked up a *Reader's Digest* to pass the time. One of the feature articles was about the Passion Play, which takes place every ten years in Oberammergau. I was astounded to read that the female lead of Mary Magdalene was none other than Krystal Rutz.

If I could choose only one place to visit, when there were not wall-to-wall people, it would be Oberammergau. The scenery is spectacular. The town is located at the foot of the Alps, with beautiful flowers, beautiful homes, and even the commercial buildings have murals painted on their sides.

In 1973, we were again in Europe. This time, the group included Ann, our son Frank, who was fifteen, and Howard and Evelyn McNeal. Once again, one of our stops was in Oberammergau.

Mrs. Rutz's Bed and Breakfast was leased out to someone else, and they could not accommodate us. But Mrs. Rutz, with her daughter Krystal, arranged for us to stay with some of their relatives in a ski resort across town. There were four ski slopes that ended at this house, and the man of the house earned his living by making and repairing skis.

Several years later, when we returned to Oberammergau, Mrs. Rutz's son was then running the Bed and Breakfast. They gave us two choice rooms. Sarah and Will were with us that time, and they too fell in love with the town. Krystal was married and living across town with her husband and two children. They invited us to dinner one evening, and Mrs. Rutz, who was still quite active, was also there.

Another enjoyable part of the trip was driving a rented car from Paris to Eyweiler, arriving there at lunchtime. As usual, we had brought along food for a picnic. Adjacent to the old schoolhouse that had served as our army mess hall back in 1945, we found a playground with a two-horse wagon.

We had no more than laid out our lunch on the flat bed of the wagon, when a lady, Erna Buschi, came out of a nearby house and insisted that we come into her dining room to eat our lunch. We declined at first, but then we finally accepted her invitation. She even added quite a few things from her kitchen.

Her home was one of the many row houses, with the attached barn and the manure pile out front. She seemed to understand a little English, although she did not speak it.

Mrs. Buschi's father, a veteran of World War I on the German side, joined us at the table, but he would have no part of our apple juice with ice. Nowhere in the rural areas of Europe do they serve drinks with ice. After lunch, the old man showed us his scrapbook, mainly from World War I. He had pictures of himself as a soldier under Kaiser Wilhelm, with pictures of the Kaiser and many of the army officers.

Mrs. Buschi showed us around her house. It was a primitive house, with an outdoor toilet, a hand-dug well, a big cellar, and a nice garden in the back, with lots of vegetables growing. The rural community had changed little from earlier days. By the time we left the Buschis, we felt as if we had visited our own grandparents.

Knowing that we would be visiting Eyweiler, we had brought a letter in German, stating who we were and why we were visiting. Because of the language barrier, this letter came in handy.

We found the house where I had spent the last two nights in 1945. It had been freshly painted and was in contrast to many of the other houses. We presented our letter to those living there, and through much sign language and patience on both sides, we learned that the people who had once owned the house had moved a short distance away.

Earlier, I had also learned that their ten-year-old son had been killed by a mine buried in the front yard. This was very odd, because many of us had walked back and forth from the road to the house. Apparently, the Germans had buried mines in the yard as they retreated. They had also left a grenade hanging in an apple tree out in the pasture. It could have been booby-trapped also, but back then in 1945, I did not touch it.

Since 1973, we have been back to see Mrs. Buschi three times. Her father had passed away and the two children we'd met on previous occasions had moved on and were working in the city. She was always very gracious, insisting on our spending the night with her, but reluctantly we would drive on to the next town.

Mrs. Buschi and I always exchange Christmas cards and other correspondence. Our spirits seem to mesh, and it is always difficult to say goodbye when visiting, since there are many tears shed. She seems exactly like a dear grandmother, who lives too far away.

When I was at Mason and Dulion Company, I worked with a plumber named Ernest Buschi. His father came from the Alsace-Lorraine area in Europe, where Eyweiler is located. Perhaps, if we could trace the genealogy, we might find some family connection between Ernest and Frau Buschi.

By 1987, we were again in Europe, but this time Ann's brother Charles, who had served in Patton's Army, and her younger sister Callie were with us.

One of our goals was to return to the American Cemetery at St. Avold in eastern France to place flowers and take a picture of the gravesite of Wayne Alderson's buddy, Red Preston, who had saved Wayne's life in World War II. The two soldiers had been in a trench together, and as a German raised his rifle to shoot Alderson, Red had stepped in front of Wayne and taken the bullet, instead. Wayne is the man that R.C. Sproul wrote about in the book, *Stronger than Steel.*

It was a rainy afternoon when we arrived, and the only others at the cemetery were two guards. As we waited for the rain to slacken, one guard and I began comparing stories. When I told him that I had been with the 121st Cavalry Reconnaissance Squadron, the guard said, "We have some of your people buried here."

When he showed me the names, I recognized that of Sergeant Tom Klump. This brought back memories of the day that his twin brother Jim and I were wounded in the mine explosion at Eyweiler.

Shortly after that event in 1945, the 106th Cavalry Group moved back north and crossed the Rhine River into the heart of Germany. On April 25, in a little town named Tagmersheim, Sergeant Tom Klump's bantam jeep was ambushed and both he and his driver were killed. Before I was wounded, I had been scheduled to become Klump's driver.

So in addition to taking a picture of Red's grave, we also took pictures of Tom Klump's. But because of the rain, all the flower vendors had gone home and we were unable to place flowers at the gravesites.

After visiting the cemetery where Tom and Red are buried, we drove on into Luxembourg and visited the American Military Cemetery at Hamm, where General George Patton is buried. Patton wanted to be interred with his troops. However, there were so many people coming to visit his grave, that it was difficult to maintain the grass and keep the site in first class condition. So Patton's grave was moved near the entrance to the cemetery, making it a lot easier for people to come by and pay their respects.

Someone had told us that we should also visit the

German Military Cemetery that was just down the road. We had a little trouble finding it, since it turned out to be farther than we were led to believe. Also, there were no markings or long rows of white crosses. Instead, the Germans were buried in a forest of trees, with one marker for every two graves. There was also a mass grave, where 5,000 were buried. What shocked us most were the dates of birth listed on the markers. Most of the German soldiers were only seventeen years old. What a waste of young manhood!

On the last leg of our trip, we drove north along the Rhine River. We wanted to stay in or near a quaint and historic little town, Rudesheim. We called ahead and found a tiny hotel across the river in Bingen.

When we arrived, the hotel dining room was closed, but the owner, an Englishman, told us of a small family-style eating place just a block away and off the beaten path for tourists. The four of us walked up a hill and around the corner. The eating place was very much like a private home. When we arrived, the only ones there were the owner and his wife, with children playing on the floor. They seated us, and the meal, as is usual with almost all German food, was excellent, as well as the beer, which they drink as we do water or iced tea.

Before we finished our meal, two men and a lady about my age came in and sat not too far from us. They were speaking German, and we were speaking English, so neither group paid much attention to the other, until one of the men, in excellent English, said, "You must be from the United States."

"Yes. How did you know?"

The man explained that he had spent six years in Canada and the USA. Interested, I said, "Tell us about it."

He told us that he had been a German soldier fighting in North Africa in 1940, before the United States got into the war. He was captured by the English and sent to a prisoner of war camp in western Canada. He evidently impressed the authorities with his trustworthiness, so they allowed him to drive trucks back and forth into the USA.

He was determined to learn as much as he could about both countries and to learn to speak good English. He remained in the prisoner of war camp until 1946, when he returned to Germany. He had recently retired from the postal service, where he had spent his entire career.

The other two people at his table had not joined in

the conversation. When I asked about the lady, the man said that she was a friend to both. Caring for her invalid mother, the only time she got any relief was on Sunday afternoons, when the two fellows took her to this restaurant for dinner.

When I asked about the other man, he said, "Oh, he was a soldier on the Eastern Front. He was at the Siege of Stalingrad." I immediately jumped up and went over to him to shake his hand. Thousands upon thousands of Germans were killed or frozen to death in minus-forty-degree weather, in that yearlong siege. Finally, the Germans retreated, and that was one of the turning points in the war.

Later, I wrote the English-speaking soldier and received a nice letter in return. After this encounter with our former enemies, Ann and I realized that Americans and Germans have more in common with each other than with some of our Allies.

Once we arrived home from this European trip and got all of our pictures developed, including the ones taken at St. Avold, the cemetery where Tom Klump was buried, I thought that Jim Klump would appreciate a picture of his brother's grave. We had had no contact with each other from the time he came by to see me in the little field hospital in France, when he was returning to his outfit. I looked in the 106th Cavalry Group's book and got the address in Haubstadt, Indiana, where Jim had lived before the war. Through a telephone operator, I was able to get his current number. He was dumbfounded after all these years to hear from me. Once he realized what I was talking about, he warmed up considerably.

He had not been back to Europe since the war and had no wish to go. Losing his twin had affected him profoundly, even though he had gone on with his life. At the time I knew him, he was an outgoing, aggressive type person, who was not afraid of anything. But he admitted that Tom's death had taken the fight out of him.

All three of us, Tom, Jim, and I, were machinists in civilian life. In 1987, Jim had recently retired from the same company where he had worked prior to the war. The company, a competitor of R.G. LeTourneau, built heavy earthmoving equipment. He had also married and had several children and grandchildren.

Since our phone visit years ago, Jim and his family made a trip to Europe to visit Tom's grave. In 1997, Ann and I visited him in his home. His wife had died, but during that same year, his sons and grandchildren spent the night with us on their way to the Talladega 500.

I have revisited Eyweiler a number of times in my European travels. It seems strange that I can remember so little of what happened to me on the day of March 5, 1945. Instead, I have had to rely on the account written in a personal letter by Eldred DeKoster, who reconstructed for me that day in the minefield.

But engraved on my memory are the houses with the manure piled in front, the gracious people, such as Mrs. Buschi and her father, and especially, the twin brothers in our Troop "B," Third Platoon, 121st Cavalry Reconnaissance Squadron.

William, David, and W.C. Stegall at Military Cemetery

Chapter 22

Golf and the Masters

The game of golf has opened a lot of doors for me. It led me to people who were very influential and helpful in my life.

Except for two or three attempts to play the game back in Longview, Texas in 1946, I had given it little attention until 1953, when I was working at Mason and Dulion Company.

Both Bob Brand and I were workaholics. Because of the long and hard hours we put in, we decided that we needed some sort of exercise or outdoor activity. I believe it was Ann, who suggested golf. So Bob and I went out to Roebuck, a golf course, and tried to play a few rounds. It was a total disaster.

On the old Atlanta Highway, near the site of Eastwood Mall, we went to Sam Byrd's to take a few lessons. We signed up for six lessons together, for a very nominal fee, like five dollars for a half-hour.

Sam had been a very successful baseball player, playing for the New York Yankees when Babe Ruth played right field. He got so interested in golf, that he gave up baseball and became a golf professional, winning one of the U.S. Opens that was held during World War II. But since it was wartime, he was never recognized as an Opens winner.

Sam helped Bob and me a lot—especially me. After we got interested and began playing, Kenny Walden, who also worked at Mason-Dulion, joined us. He had played golf some years back, so every Saturday, the three of us played at either Highland Park or Roebuck. Since we had to work until noon on Saturdays, we would often try to slip away a little early to get an early tee time. Mr. Dulion's grandson, Nick Vance, had loaned me a very good set of clubs, which, according to young Nick, I wore out.

Back in 1953 or 1954, getting tickets to the Masters Tournament in Augusta, Georgia was no problem. You could actually walk up to the gate and buy daily tickets. And finding a place to stay was no problem, either. Many residents opened up their homes for as little as one dollar per night.

Watching such great golfers as Hogan, Snead, and DeMaret, plus all those who have since come along, really whetted our appetites for golf. I have gone to the Masters almost every year from 1953 to the present. Gene Burks, a salesman for Moore-Handley, Inc., went with Bob and me several times. I remember rooming with Gene at a private residence, and I was surprised to see him shaving with a straight razor, as well as wearing thermal underwear. But as everyone knows, the weather can turn cold at the Masters.

After a few years, acquiring tickets grew scarcer. I was offered the chance to pre-purchase eight badges, which I did for years. Finally, the number was reduced to four.

Sometimes, four of us, and then sometimes eight of us traveled to Augusta, without a place to stay. We always managed to get a soft bed, although we ran into a few comical situations.

On one occasion, my good friend, Eddie Allen,

with Daniel Construction Company, and I went, just the two of us, with no place to stay. We finally found an accommodation on the outskirts of Aiken, South Carolina. It was in a nursing home for elderly ladies, with one room available. It was really cold that night, and the only heat came from a potbellied stove, which we fired up. Soon, the fire got so hot that we had to open all the windows. But we lived through it.

On another occasion, George Jackins, a good and personal friend, and I decided to go at the last moment. George had learned to fly, and had just gotten his license. He went out to the local airport and rented a little puddle jumper of a plane. Early in the morning, we took off for Augusta.

The old airport had opened up to accommodate private planes coming in for the tournament. A tower operator had been engaged for the weekend. But this was Thursday, so it was every man for himself when he was ready to land. With all the planes coming in, it reminded me of a beehive on a hot summer day, with all of the bees coming and going.

We landed safely and watched golf all day. At about five o'clock, when the playing stopped, we began to wonder where we would spend the night. I told George not to worry. We could simply get into the airplane and fly to the nearest town with an airport. Finding a room in a nearby South Carolina town should not present a problem. At the worst, we could go to Columbia, which wasn't that far.

We got into the plane and headed east. Not long afterwards, we began to get blasted on the radio, since we were flying over the Savannah River Nuclear Plant Project, which was absolutely off limits to all private aircraft. We detoured immediately, without getting shot down.

The sun was still up, so I suggested to George that we might as well fly on to Charleston and spend the night. We would have an interesting time and enjoy some fine food. This suited George just fine.

As we neared Charleston, I was trying to read the map to get the correct radio frequency for the airport. We flew and flew. Then, I realized that what I thought was the number 8 was actually a 0, with a river running through the middle of the number.

About that time, a U.S. Navy jet started buzzing all around us. It seems that we were right on top of the

Naval Air Station in Charleston. We finally zeroed in on the civilian airport and made a safe landing. From the airport, we took a cab downtown, to the heart of Charleston, where we found an old hotel at the end of the Battery. There, we got a nice room and an outstanding meal, including she-crab soup.

For years, when various friends or members of the family went to the Masters, we stayed at the Firecracker Motel in Warrington, Georgia. It was not a fancy place, but it was adequate, and it was often filled with people from Birmingham. Later, when it changed ownership, we found another place.

For the last several years, except for 1998, we stayed at the Inn on Main in Edgefield, South Carolina, twenty-two miles from Augusta. This was really a rustic place in a very historic town. The drive back and forth became a little too much, although the beautiful farmland we passed through was quite scenic.

In 1997, Ann and I stayed in Edgefield for four nights, when Charlie and Marion Wylie, of Northern Ireland were our guests. The eighty-one-year-old Charlie had been an avid golfer all his life, as well as an outstanding evangelical. We picked them up in Atlanta on the

Monday of the Masters week and drove on to Edgefield.

I wanted to take Charlie and Marion to a practice round where we could take pictures, talk to the players, and have freedom of the course. But try as hard as I might, I could not get practice tickets, even for one day.

Finally, I decided to call on my good friend, Ken Venturi, who for the past seventeen years had been coming to Shoal Creek, where I live, to help entertain Rust's Engineers special customers from all over the world.

Rust had invited me to each of these functions, since they needed a Shoal Creek club member in each foursome. Venturi and I got fairly well acquainted during all this time. And when I saw him at the Masters, he always had a friendly word.

That previous October, before the Masters Tournament in April, I had asked Venturi's help in my getting four practice round tickets for Tuesday only. He said that would be no problem, and gave me instructions on how to get in touch with him when I got to Augusta.

Following his instructions, I had Ann and the Wylies drop me off at the CBS gate, which was down the road from the main entrance. Then Ann parked our car in the special parking place. The guards, who were manning the CBS entrance, were not especially friendly. I told them that I was merely following instructions to come to this entrance and ask for a guard named Billy. The four or five people standing around looked at each other in a puzzled way, as if they had never heard of Billy. Finally, one of the guards broke out laughing. He was Billy. When I gave him the instructions from Venturi, he called over a young man with a golf cart and he took me down to the CBS complex.

Venturi was drinking coffee in his office. At first, he acted surprised to see me. Then he finally remembered that he had promised to get me tickets, but instead of four, he came up with only two, one daily pass and one CBS badge. Explaining that the only reason I had asked his help for four tickets was because of our special guests from Northern Ireland. He pretended to be a little irritated when he went into the Program Director's office. His boss evidently went along with the humor, fussing at both Venturi and me about the impossibility of getting four tickets. Then the director rambled through his desk and brought out two CBS badges, with CBS personnel names on them. He told Venturi and me that I could have those badges for the day, but that if they were not returned by

5 P.M. that afternoon, he was going to shoot Venturi first, and then come after me. They both seemed to have fun giving me a hard time.

The young boy who had waited with the golf cart wound through the woods as he took me to the main entrance, where Ann and the Wylies were waiting. We put on our CBS badges, which later caused quite a stir. We were given much deference and plenty of room all day. It was a neat experience, not only for the Wylies, but for the Stegalls, too. And we got the badges back by 5.

Several years before that incident, Venturi had told me when he was at Shoal Creek that if I would contact him on an early Thursday morning of the Masters, long before the golf telecast began, that he would give me an inside view of how a golf tournament was televised. That year, Delbert Bright and his wife Jean were our guests. Delbert and I went together to the CBS headquarters, located far back in the woods. We were tremendously impressed with all the technology, instrumentation, and equipment.

Delbert Bright's golf pro back in Longview, Texas had been using the instruction manual that Venturi had produced, and it showed its years of use. The young pro wanted Bright to get Venturi to autograph it for him. Bright was a little hesitant to ask, but when I realized what he wanted, I immediately asked Venturi if he would sign it. When he discovered that someone had used his manual all these years, Venturi seemed honored. The next time I was in Longview, the young pro insisted on giving me free lessons, using the TV pictures of my swing, and pointing out my good and bad points.

For the past several years, a series badge covering four days at the Masters costs $100, with the black market price as high as $6,000. Yet, a person does not even need to play golf in order to enjoy and appreciate the Masters. I know of no other place in the world that is more beautiful during the first week in April. The azaleas, dogwoods, and various and sundry plants are breathtaking.

One of the greatest enjoyments of going to the Masters is to go with someone who has never been. People cannot believe how beautiful it is.

During the 1999 Masters, eighteen of the Stegall family used the badges, some for only a half day. I wish I had kept a record of all the friends, as well, that we have let use our badges through the years. In no particular order or particular year, here are a few of them:

Bob Brand, Gene Burks, Eddie Allen, Stan Simpson, Jim Judd, Ben Sibley, George Jackins, Bill Watson, Jack and Nelva Ballard, Jim and Betty Mulvaney, Jack and Carol Mayhall, William and Mary Stegall, David and Deborah Stegall, Will Stegall, George Stegall, Frank Stegall, Ben Lincoln, Homer Lincoln, Jim and Beverly Brooks, Charles Williams, Delbert and Jean Bright, Jimmy and Sarah Bragan, Billy and Libby Spain, Fred and Jennie Star (they helped us play host on several occasions), Stephen and Heather Alford, Sumner Wemp, Walter Wood, Harold and Barbara Hunt, Harold Coffey, Weedie Stegall (mother), Charlie and Marion Wylie, Wink and Grace Smith, Bobby and Roxanne Nix, Ammon and Barbara Stoltzfus, Bud Newbold, Fred Threadgill, George Welden, Mr. Tatum, friend of Fred Star, Wayne and Alice Cartee, David Cartee, Burt and Leigh Ann Young, James Mulvaney, K.C. Squires, Sarah Stegall Squires, Emalee Stegall, Jessica Young, Frank Young, Caleb Young, Hillary Stegall, Stephen Perlis, Spencer Stegall, Taylor Stegall, and Matt Stegall.

Chapter 23

TESTAMENT

The most pleasant part of growing up on a farm is "laying by" time, from the middle of July, when the last plowing of crops occurs, to September 1, with the harvesting of crops, such as cotton and corn.

When I was a boy, these six weeks were spent playing baseball, fishing, and for two weeks straight, attending religious revivals, or protracted meetings, which were the highlights of the social year, especially for all of the young people.

Our neighbor down the road would hitch up his two-horse wagon, fill the body with straw, help his wife and children climb aboard, and then stop by our house to pick up our family. Our old hound dog, as well as our neighbor's, would always follow behind the wagon on the way to the revival meeting.

Preaching at the local Baptist Church took place each morning, before the sun got hot, and each evening after sundown, in the cool of the evening. I can recall the old Baptist preacher really laying it on. I was so stirred that I really wanted to go with the others and give my heart to God. There was just one problem. All my past sins would be forgiven, but from then on, it was my understanding that I would have to live a perfect life. Since I knew this was not possible, I just held on to the wooden bench harder and harder, until finally the services were over for the year.

My next spiritual enlightenment came when I was a student at Rabun Gap-Nacoochee Junior College. It was a Christian school, where all students took a class in Bible. Several teachers and students had an important influence on me, but the one who stood out spiritually above all was Miss Alice Lennon. She was seventy-five years old, six feet tall, and weighed about one hundred and ten pounds. She also could recite nearly all of the New Testament, whereas I could not even recite John 3:16.

She started every Bible class with prayer, and in

her Bible survey course, she covered the entire Bible. I still have, and often use, the outline form all her students had to prepare.

The summer after graduation from the junior college, I played baseball for a textile team near Anderson, South Carolina. Even though I loved baseball, I just could not take the life style of those around me. I guess that, being overly protected at home and school, I had no idea that such things went on in the world. Before the baseball season was over, I learned that my former coach, Jack Acree, and many of my old buddies at Rabun Gap had entered a training program at LeTourneau, an industrial plant near Toccoa, Georgia.

On July 29, 1941, when I was off from work, I went to Toccoa, and it was only hours before I had signed up for the same training program. At that time, it was called the LeTourneau Machinist School, a two-year program where we worked eight hours a day and went to school at night. About seventy-five of us lived in barracks-type dormitories, with twenty-five working on each of the three shifts. Jack Acree was our leader, except for the shop work.

One day in October of that year, I stayed at the

shop a little longer than the others and was walking alone on the little dirt road to the dormitory. A brand new Oldsmobile came alongside, and a very attractive and well-dressed lady offered me a ride. I explained to her that I was only going a few hundred yards more and, besides, I was too dirty and greasy to get into her new car.

After insisting that I get in, since she was going to the dormitory anyway, I accepted her invitation. She then introduced herself as Mrs. R.G. LeTourneau. I was astonished that the wife of the founder and president of this huge industrial enterprise would even notice me, much less stop and offer me a ride. A wonderful relationship with "Mom" LeTourneau began that day, and continued for over thirty-eight years.

Several weeks later, I had a bad case of flu that really knocked me out. I smoked back in those days, constantly had colds, and usually woke in the mornings, coughing. When Mom LeTourneau found out that I was not at work, she drove to the dormitory and took me to her home, a very beautiful place on the shores of Lake Louise, near the plant.

At the time, Mom had four boys of her own, but she treated me just like one of them. She kept me there for several days, feeding and doctoring me, and showing real concern for me. I do not remember her preaching to me, but I do remember later on that she said it was not her natural self that caused her to have concern for the seventy-five boys who worked at the plant. It was a result of Christ living within her.

During the last week in October 1941, a citywide revival was held at the Toccoa High School auditorium. Mom rounded up all the boys she could and took us to this revival. Jesse Henley, a Baptist preacher from Atlanta, led the revival. As the invitation was given, I went forward. I believe the decision had been made in my heart to accept Jesus as my Savior, and this occasion was only to make it public and official.

I remember that later that evening and during the days to follow, I did not feel any different. In a way, I was a little disappointed because, at least subconsciously, I had expected a big emotional experience.

Those of us who became Christians during these days attended the First Presbyterian Church in Toccoa. It was an old historic church, with Dr. Forrest as pastor. He was the founder and president of Toccoa Falls Institute for many years. He was also a good friend with R.G.

LeTourneau, and to a great extent, it was his influence on R.G. that caused him to locate a plant in Toccoa. At that time, the main LeTourneau plant was in Peoria, Illinois.

During the summer of 1942, the U.S. government wanted LeTourneau to build more and more machinery and to begin to build 155mm shell casings. About twenty-five of us, who had been with the Toccoa program for a year, were transferred to the new plant in Vicksburg, Mississippi.

Before we went to Vicksburg, Mom took a number of us to Winona Lake, Indiana, for a week's vacation. Winona Lake, then, was more or less the Christian camping headquarters of the nation. This was quite an experience for an old country boy. One of the counselors at the camp that summer was a young man about my age, Billy Graham. I do not remember him from that time. After he had become famous, Mom LeTourneau reminded us that we had been at the Christian camp with him.

In Vicksburg, I joined a Baptist church, and was baptized there. It was a good thing, since Vicksburg, at that time, was a wild riverfront town, with temptations of every sort. It was almost more than a young man could handle. If it had not been for those months of spiritual nurture before arriving there, I shudder to think what could have happened.

By 1944, all able-bodied men were needed in the army. The companies making weapons of war could no longer get deferments for their workers. My newfound faith was a source of comfort during exhaustive combat training and later in Europe, when I was wounded.

After the war, when my dear friend Ann and I were married, I was still a member of Bomar Baptist Church in Vicksburg, Mississippi. Ann and her family had been Methodists for generations. In fact, her brother, a Methodist minister, performed our wedding ceremony.

My own introduction to the Methodist church came after we moved to Seneca, South Carolina, where we joined the Seneca Methodist Church. During the years that I was a student at Clemson, we both attended the Methodist church on campus.

In the first part of our marriage, Ann and I fussed about money more than anything else. This went on for several months until finally the Lord began working on us individually. Together, we decided to tithe our income. I can sincerely say that we have had no problems concerning money since then. That doesn't mean that we have had all the money we would like, but since that date, we have had no more fusses, nor any real anxieties over money.

R.G. LeTourneau used to have a favorite scripture: "Seek ye first the kingdom of God, and all these things shall be added unto you."

After graduation, we moved to Philadelphia, again affiliating with the Methodist denomination. But a few months later, we were in New Jersey. Here, we had very little church life. We attended a Congregational church a few times. It was when we made the move to Birmingham that we really became actively involved in church life.

We were members of Canterbury Methodist Church until we moved to the East Lake area and consequently joined a church close by. East Lake Methodist's congregation was extremely large. Ann worked with the little ones and in her ladies' Sunday School class, while I often taught the senior high school students and occasionally substituted in the George Turner class. And it was here that I first became a steward, now known as a member of the Official Board.

By 1958, we had moved to Vestavia Hills and joined a young, growing church in the vicinity. We got to know Bud and Elizabeth Newbold at the Vestavia Hills

Methodist Church, and I became Bud's assistant, as a teacher. At other times, I taught the college age group and again was elected as a member of the Official Board.

During this time, Ann and I, along with the Newbolds, often went to the Christian and Missionary Alliance Church, located in the middle of Birmingham and across the street from City Hall. The evangelical sermons spoke to our hearts and souls, and strengthened us spiritually.

In the 1960s, Bill and Janie Buck started a Bible class at their home on Vestavia Drive. Elizabeth Newbold did most of the teaching, with her husband Bud occasionally teaching also. A personal testimony became a part of the Bible class each time. This class, with the teaching and testimony, had tremendous effect on perhaps hundreds of people, who gained assurance of the richness of salvation.

Ann and I attended when we could, but with four children who needed our attention, we were not regular participants. Yet, it was at one of these Bible classes that we first met Frank Barker, who introduced us to the Briarwood Presbyterian Church. Until then, we had not known of this new little church being established in the Cahaba Heights area.

The Bible class began to grow, until the house could not hold the two hundred people who were showing up. It had been such a spiritual success that the class had to be abandoned, for lack of space. That became a blessing, however; because smaller Bible classes sprang up all over the city. The seed, like the mustard seed, had been planted at the Bucks.

Another result of that original Bible class is that many who attended became members of Briarwood Presbyterian Church. While Ann and I were still members of Vestavia Methodist, we went to Sunday School at Vestavia, and then to morning church services at Briarwood. The evening services there were very informal and enjoyable, so we attended as many of those as we could. The Newbolds joined Briarwood on the last Sunday of 1967, and the Stegall family joined on the first Sunday of 1968. Neither family knew of the other's intentions. Our entire family found a home there and received a tremendous blessing. For thirty years, until 1997, Ann and I were a part of Briarwood.

We both served in responsible positions—Sunday School teacher, Bible class teacher, chairman of Christian School Board, chairman of Home Missions Committee,

Ministerial Aid Committee, deacon, elder, and trustee. Ann opened our home and played hostess to many groups and individuals.

From the pulpit one Sunday, Frank Barker announced that a new little church, the Presbyterian Church of the Hills, was being established and the church needed a nucleus of people who could go over and help them get started. He suggested that some of our members do that for six months or more.

Ann and I went, and on the first Sunday we both fell in love with the group. We already knew many of the people, and the church had such a friendly atmosphere, acting as if it really wanted and needed us. On December 1, 1997, Ann and I joined. I was elected an elder, and I have taught the adult Sunday School class on several Sundays, and hope to participate more, as needed. We believe that this is where the Lord would have us stay, although we still maintain friendly relations with everyone involved at Briarwood.

We have been surrounded by outstanding Christian friends—the officers and staff of Briarwood Church, Campus Crusade for Christ, Navigators, the Keswick Ministry, Salvation Army, and the various Bible and Sunday School classes.

For many years, I have had as a goal to get to the point in my life where I cannot distinguish whether or not I am plowing, playing, or praying. The first work that a farm boy does is to plow, so to me, plowing represents work or business. The playing represents recreation, golf, or what we normally consider to be fun and entertaining. Praying, of course, represents one's communion with the Lord, by Bible study, worship, and being at the right place in life at the right time. Almost every activity of life can and should contain all three elements. There have been a few times in my own life when these three have come together in perfect harmony.

My spiritual journey from the protracted revival meetings near Bear Swamp, when I was a child, to the present day, has been one of learning, of faith, and of assurance that, no matter how rugged the journey, we are never alone on our ride to Glory.

Wedding Day, December 29, 1946
Ann Welden and William Clyde Stegall

Golden Wedding Anniversary, December 29, 1996
Ann and William Clyde Stegall (top step) with Family

Editor's Note

"Let me tell you a little story."

Those were the first words that Bill Stegall said when I first met him in 1998.

I had driven from Atlanta to Birmingham to interview him for a chapter in the book, *Mountain Legacy,* a story of Rabun Gap-Nacoochee School. The book had been commissioned as a part of the centennial celebration of the mountain school that had given so many of its students a chance to earn an education. Known to his friends simply as "Stegall," he was one of its outstanding alumni.

Warm, imposing, and with a wonderful sense of

humor, this six foot, six inch man, I discovered, had been telling stories all his life. And those stories deserved more than one chapter in a book.

Shortly after that interview, Stegall discovered that he had cancer. Perhaps that was why he was so intent on finishing what he had begun—the story of his life from that small boy growing up in Bear Swamp to the present day.

Between chemotherapy treatments, Bill Stegall gathered his fragments of memory and set them down on paper. Several months later, I received his papers in the mail and almost the next day, he was gone.

He made no attempt to disguise the poverty and violence of growing up during the Great Depression, or the shortcomings of some of the people he loved. He was searching for truth, and his stories are the better for it.

Frances Patton Statham, Editor